MW01253026

Dynamics of Entry and Market Evolution

Also by Jati Sengupta

COMPETITION AND GROWTH: Innovations and Selection in Industry Evolution

INDIA'S ECONOMIC GROWTH: A Strategy for the New Economy

Dynamics of Entry and Market Evolution

Jati Sengupta
University of California, Santa Barbara, USA

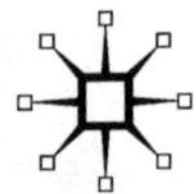

First published 2007 by
PALGRAVE MACMILLAN
Houndmills, Basingstoke, Hampshire RG21 6XS and
175 Fifth Avenue, New York, N.Y. 10010
Companies and representatives throughout the world

PALGRAVE MACMILLAN is the global academic imprint of the Palgrave Macmillan division of St. Martin's Press, LLC and of Palgrave Macmillan Ltd. Macmillan® is a registered trademark in the United States, United Kingdom and other countries. Palgrave is a registered trademark in the European Union and other countries.

ISBN-13: 978–0–230–52153–7 hardback
ISBN-10: 0–230–52153–3 hardback

This book is printed on paper suitable for recycling and made from fully managed and sustained forest sources. Logging, pulping and manufacturing processes are expected to conform to the environmental regulations of the country of origin.

A catalogue record for this book is available from the British Library.

Library of Congress Cataloging-in-Publication Data
Sengupta, Jatikumar.
Dynamics of entry and market evolution/Jati Sengupta.
p. cm.
Includes bibliographical references and index.
Contents: Entry dynamics: theory and implications — Innovation and efficiency in industry evolution — The costs and effects of market entry — Entry and market structure — Industry evolution under entry barriers — Model of industry evolution under innovation.
ISBN-10: 0–230–52153–3 (cloth)
ISBN-13: 978–0–230–52153–7 (cloth)
1. Production functions (Economic theory) 2. Corporations—Growth—Mathematical models. 3. Industrial efficiency—Mathematical models. 4. Technological innovations—Economic aspects. 5. Industries—Technological innovations. I. Title.
HB241.S37172007
338′.064—dc22 2006052509

10 9 8 7 6 5 4 3 2 1
16 15 14 13 12 11 10 09 08 07

Printed and bound in Great Britain by
Antony Rowe Ltd, Chippenham and Eastbourne

To Ramakrishna and Sarada

Contents

List of Tables

Preface

This book discusses why firms grow and decline. How does competition affect this process? Industry evolution today depends critically on innovations and R&D investments. The book analyzes the theory of Schumpeterian innovations in many forms and its impact on the selection and adjustment process in industry evolution.

Both Walrasian and non-Walrasian adjustment mechanisms in evolution are discussed here in terms of core competence and efficiency theory. The stochastic aspects of the entry and exit process and the nonparametric treatment of the R&D externalities provide some new insights. The book emphasizes the applied and empirical aspects of evolutionary dynamics and as a case study the computer industry is analyzed in some detail over the years 1985–2000 in respect of innovation efficiency, learning by doing and the R&D spillover effect through demand growth.

Finally, I wish to record my deepest appreciation to my two great teachers: to my Guru for his advice to lead a life dedicated to the Divine and to my father who taught me to always seek inner bliss in life.

Jati K. Sengupta

1 Entry Dynamics: Theory and Implications

1.1 Introduction

Entry and exit behavior of firms tends to determine the pattern of industry evolution. Industries grow when new firms enter the industry; they decline when the incumbent firms exit. The entry process necessarily involves competition between the incumbent and the new entrants. Models of perfect competition where firms are price takers, assume that entry and exit are more or less costless and under conditions of free flow of market information this yields an industry equilibrium. The Walrasian model of competitive equilibrium introduces two types of adjustments: price–cost adjustment and demand–supply adjustment. First, if price exceeds marginal cost where the latter could be viewed as minimal average cost, the profitability continues and this provides incentives for new firms to enter the market till the excess profits are eliminated by new entry. Secondly, if excess demand persists and the capacity ceiling is hit, prices tend to rise thus causing profits to rise. This again provides incentives for new entry or the expansion of output by the existing incumbents.

We consider here selected models of entry dynamics and their implications for industry evolution. Competitive models are first selected and then followed by game-theoretic models.

1.2 Competitive models with entry

Here we assume that there are no costs of entry, so that the competitive incentive to entry is due to sustained profitability in the industry.

The entry model due to Veloce and Zellner (1985) considers the entry equation in a linear form as

$$\frac{\dot{N}}{N} = \gamma(F_e - \pi), \quad \gamma > 0 \tag{1.1}$$

where N is the number of firms, dot is the time derivative, π is aggregate profits and $F_e = F_e(t)$ is aggregate long-run profits which may include entry costs so long as these are small and competitive. If $\pi < F_e$ there is positive entry, while $\pi > F_e$ represents exit or negative entry. Assuming two inputs labor (L) and capital (K) and one homogeneous output q with a Cobb–Douglas production function

$$q = AL^{\alpha} K^{\beta}, \; 0 < \alpha + \beta < 1; \; \alpha > 0, \; \beta > 0$$

an individual firm's profit π is

$$\pi = pq - wL - rK$$

where p, w, and r are the prices of output and the two inputs. The optimum profits are then $\pi = (1 - \alpha - \beta)s$; $s = pq$ is sales.

The firm's supply function (s) and the industry supply function (S) can then be written as

$$\begin{aligned} s &= A_1 p^{1/\theta} w^{-\alpha/\theta} r^{-\beta/\theta}, \quad \theta = 1 - \alpha - \beta > 0 \\ S &= Ns = NA_1 p^{1/\theta} w^{-\alpha/\theta} r^{-\beta/\theta} \end{aligned} \tag{1.2}$$

On taking logarithms and differentiation with respect to time t yields

$$\frac{\dot{S}}{S} = \frac{\dot{N}}{N} + \left(\frac{1}{\theta}\right)\frac{\dot{p}}{p} - \left(\frac{\alpha}{\theta}\right)\frac{\dot{w}}{w} - \left(\frac{\beta}{\theta}\right)\frac{\dot{r}}{r}$$

The industry demand function is assumed to be of the Cobb–Douglas form

$$Q = Bp^{-\varepsilon} x_0^{\varepsilon_0}$$

where Q is aggregate demand, $-\varepsilon$ is the price elasticity of demand and x_0 is a composite variable representing other variables like income, etc. They express the demand function Q in terms of industry sales $S = pQ$ as

$$S = Bp^{1-\varepsilon} x_0^{\varepsilon_0}$$

or

$$\frac{\dot{S}}{S} = (1-\varepsilon)\left(\frac{\dot{p}}{p}\right) + \varepsilon_0\left(\frac{\dot{x}_0}{x_0}\right)$$

Since $\Pi = N\pi = (1-\alpha-\beta)$ Ns $= \theta S$ one can write the entry equation as

$$\frac{\dot{N}}{N} = \gamma(F_e - \theta S) \tag{1.3}$$

Thus Veloce and Zellner obtain the three-equation competitive entry model (CEM)

1. Demand $\dot{S}/S = (1-\varepsilon)(\dot{p}/p) + \varepsilon_0(\dot{x}_0/x_0)$
2. Supply $\dot{S}/S = \dot{N}/N + (1/\theta)(\dot{p}/p) - (\alpha/\theta)(\dot{w}/w) - (\beta/\theta)\,\dot{r}/r$
3. Entry $\dot{N}/N = \gamma(F_e - \theta S)$ with $0 < \theta < 1$

By combining all these equations the optimal path of equilibrium industry sales in terms of the nonlinear differential equation for $S = Ns$ can be obtained:

$$\frac{\dot{S}}{S} = a\left(1 - \frac{S}{F}\right) + g \tag{1.4}$$

where

$$a = -\gamma\theta^2 \frac{F(1-\varepsilon)}{[1-\theta(1-\varepsilon)]}$$

$$F = \frac{F_e}{\theta}$$

$$g = \frac{\left[(1-\varepsilon)\left(\frac{\alpha\dot{w}}{w} + \frac{\beta\dot{r}}{r}\right) + \varepsilon_0\frac{\dot{x}_0}{x_0}\right]}{[1-\theta(1-\varepsilon)]}.$$

Two important implications follow. First, if a is positive as is most likely, then the stationary solution $S = F$ is stable, since for $S > F$, $\dot{S} < 0$ and for $S < F$, $\dot{S} > 0$. This implies that as the aggregate long-run profits F_e rise, it leads to increased industry sales S. Secondly, for given F the solution of the homogeneous differential equation for industry sales can be explicitly computed as

$$S(t) = \frac{F}{\left[1 - \left\{\frac{1-F}{S(0)}\right\} e^{-at}\right]}, \quad a,\ F > 0$$

which is a form of the logistic function with $S(0)$ as industry sales at time zero. The entry equation can now be written as

$$\frac{\dot{N}}{N} = \gamma(F_e - \theta S(t)) \tag{1.5}$$

In addition if we ignore the other variables x_0 in the demand equation, we obtain the time path of prices given by

$$\frac{\dot{p}}{p} = (1-\varepsilon)^{-1}\frac{\dot{S}}{S}$$

Veloce and Zellner have provided some empirical estimates of the supply and entry relations for the Canadian household furniture industry over the period 1959–1981. With y_t denoting $\ln Q_t$ = natural log of annual real value of shipments, w_t as real wage rate per hour and r_t as real rate of interest measured by the difference between the nominal rate on 10-year Canadian industrial bonds and the rate of change of Canadian CPI, the industry supply function estimated by ordinary least squares (OLS) appeared as follows:

$$y_t = \underset{(5.93)}{-10.1} + \underset{(1.33)}{3.42} \ln p_t + \underset{(0.012)}{0.058}t - \underset{(0.40)}{0.13} \ln w_t - \underset{(0.17)}{0.35} \ln r_t$$
$$+1.0 \ln N_t \; (\hat{\sigma} = 0.077)$$

Note that the form of the supply function (1.2) implies that the coefficient of $\ln N_t$ is one. The standard error of coefficients is given in parentheses. When the entry variable is dropped, the estimates of the coefficient of $\ln w_t$ change to a positive value 0.49 with a standard error of 0.38. This is contrary to theoretical expectations.

For the entry equation let y_t be $\Delta \ln N_t$ then the estimate turned out to be

$$y_t = \underset{(0.008)}{0.0019} + \underset{(0.17)}{0.95}\, y_{t-1} - \underset{(0.17)}{0.65}\, y_{t-2} - \underset{(0.095)}{0.205} \ln r_{t-1}$$
$$(\hat{\sigma} = 0.038, \; \bar{R}^2 = 0.6115)$$

Clearly the trend coefficient 0.95 in the entry equation is highly significant and the real rate of interest has a significant effect on entry behavior. The industry supply equation has a highly positive elasticity coefficient (i.e., 3.42) and the effect of the time trend variable is also very significant in a statistical sense.

Note, however, that Veloce and Zellner did not estimate the entry equation of the form (1.5), since data on F_e were not available.

Now we consider other competitive models of entry. In his book *Barriers to New Competition* Bain (1956) discusses three main reasons for barriers to entry: absolute cost advantages of the incumbent firms, economies of scale and product differentiation advantages. One reason for absolute cost advantages is superior efficiency of the incumbent firms. The higher profit accruing to the more efficient among incumbent firms can be considered an efficiency rent. These are the firms with more inputs of a scarce factor which includes special entrepreneurial talent. In such a situation the marginal incumbent firm will receive in equilibrium a normal rate of return on its investment.

Investment (u) by new entrants can generate scale economies in a competitive industry. Hence the growth in number of firms may be modeled as

$$\dot{n} = a(\bar{c} - c(u,n)), \quad a > 0$$

where $\bar{c}$ is industry-wide average cost and $c(u,n)$ is the average cost of new entrant which is assumed to depend on investment and the number of firms. If $c(u,n) < \bar{c}$, then there is positive entry. The growth of industry average costs depends on the levels of $\bar{c}$ and n, that is,

$$\dot{\bar{c}} = b_1\bar{c} - b_2 n, \quad b_1,\ b_2 > 0$$

Since the number of firms is not continuous, one can write the above system in terms of sales $S = nq$ as

$$\begin{aligned} \dot{S} &= a(A - c) \\ \dot{A} &= b_1 A - b_2 S \end{aligned} \tag{1.6}$$

where $A = \bar{c}$ and $c = c(u,n)$ and it is assumed that new entry tends to reduce average industry costs. On combining the two equations in (1.6) one obtains a second-order linear differential equation as

$$\ddot{A} - b_1\bar{A} + ab_2 A - ab_2 c = 0 \tag{1.7}$$

This specifies a time path of industry average cost. When optimal average costs $c = c(u,n)$ fall due to increased investment u, the new entrants increase their output. This leads to increased industry sales through higher demand. Increased S causes industry average costs to decline, thus closing the gap between A and c. The time path of the entry equation (1.7) may then be explicitly computed in terms of its characteristic roots:

$$\lambda = \left(\frac{1}{2}\right)(b_1 \pm (b_1^2 - 4ab_2)^{1/2}) = \left(b_1 - \frac{\theta}{2}, \frac{\theta}{2}\right)$$

when $(b_1^2 - 4ab_2)^{1/2} = b_1 - \theta,\ \theta > 0$. Clearly for $\theta < \mathrm{b}_1$ the two roots are positive and the equilibrium is an unstable node.

Note that the entry process here does not incorporate specifically any entry costs. However, with low entry costs, competitive process may still encourage entry and exits from the industry. Folster and Trofimov (1997) consider such a model, where the entry equation is written in terms of the number of firms n as

$$\dot{n} = \alpha(v - z) \tag{1.8}$$

with z as entry costs and v is the incumbent firm's value at time t

$$v(t) = \int_t^\infty \mathrm{e}^{-r(\tau-t)}\ \pi(n,u)\ \mathrm{d}\tau$$

with π as the discount rate. On differentiating $v(t)$ one obtains

$$\dot{v} = rv - \pi(n,u)$$

where $\pi = \pi(n,u)$ is expected net profits of a representative incumbent firm with u as the R&D parameter, i.e., it is the effort made by the firm in product innovation at time t. By assuming that u is optimally chosen one could write $\pi(n) = \max \pi(n,u)$, where $\pi(n,u)$ shows decreasing returns to R&D at the firm level. Thus we obtain the entry dynamics as

$$\begin{aligned} \dot{r} &= \alpha(v - z) \\ \dot{v} &= rv - \pi(n) \end{aligned} \tag{1.9}$$

Since n is not continuous one may replace it by sales $S = nq$. This yields the transformed system

$$\begin{aligned} \dot{S} &= \alpha(v - z) \\ \dot{v} &= rv - \pi(S) \end{aligned} \tag{1.10}$$

This model (1.10) assumes that the instantaneous cost of entry and exit is a linear function of the absolute change in number of firms. The economic interpretation of entry cost according to Folster and Trofimov (1997) is that when more firms enter they temporarily increase demand for specialized labor and capital input services faster than supply. Clearly the long-run value of the firm is $v^* = z$ in cases when it exists. When $v = z$ there is no entry or exit in the long run. This yields the steady state number n^* of firms which can be derived from the equation $\pi(n) = rz$. Now consider the solution of the dynamic system (1.10). It has the characteristic equation

$$\lambda^2 - r\lambda + \alpha\pi'(n) = 0, \quad \pi'(n) = \frac{\partial \pi}{\partial n}$$

with two roots

$$\lambda = \left(\frac{1}{2}\right)\left\{r \pm (r^2 - 4\alpha\pi'(n))^{1/2}\right\}$$

Normally we would have $\tau'(n)$ negative and in this case the two roots are real with opposite signs. In this case the stationary state (v^*, n^*) is a saddle point equilibrium, which implies that there is a stable manifold along which the industry evolution is purely towards (v^*, n^*) and an unstable manifold where the evolution is exclusively away from the equilibrium.

When $\pi'(n^*)$ is positive and α is sufficiently small, both characteristic roots are positive and the medium level equilibrium is an unstable node. If $\pi'(n^*) > 0$ but α is large, the roots are complex valued with positive real parts and this case exhibits an unstable focus. Here entrants' adjustment costs are very negligible, since α is large. The dynamic system (1.10) then exhibits explosive cyclical fluctuations. The cyclical dynamics imply a multiplicity of equilibrium paths when $\pi(n)$ is nonlinear.

1.3 Capacity investment in Cournot–Nash equilibria

We now consider noncompetitive market structures and explain how these structures evolve over time and perhaps ultimately converge to a stationary state. Reynolds (1987) considers a two-player differential game model which explains the preemptive effect of capacity investment in an environment in which investment is reversible but capacity is subject to adjustment costs. This model is closely related to the work of Dixit (1980), Spence (1979) and Fudenberg and Tirole (1983), who analyze the commitment value of capital investment for increasing capacity. In Dixit's model the incumbent firm selects a capacity level prior to the play of the post-entry game, which helps the firm to manipulate the initial conditions so as to secure a greater post-entry output. Spence (1979) analyzes the implications of a first-mover advantage and strategic investment for a duopoly firm, when investment is completely irreversible (i.e., net investment is nonnegative). The irreversibility helps the incumbent firm to preempt investment by its rival and attain a long-run market share and profitability advantage. Fudenberg and Tirole (1983) show that such preemptive investment can be part of a subgame perfect Nash equilibrium (NE) for the Spence model. In their analysis a subgame of the complete game forms a dynamic game that begins with any initial capital stock level.

Reynold's dynamic differential game assumes reversible investment with capacity changes subject to adjustment costs. Each firm is assumed to maximize the discounted present value of its cash flow stream:

$$\Pi_i = \int_0^{\infty} e^{-rt}[R_i(K_i) - C(I_i)]\, dt \qquad (1.11)$$

where investment cost is assumed to be strictly convex:

$$C(I_i) = qI_i + \frac{cI_i^2}{2}, \quad q > 0, \quad c > 0, \quad i = 1, 2 \qquad (1.12)$$

where q is the unit cost of acquisition of investment goods and c is the adjustment cost parameter. The capacity $K_i(t)$ follows the dynamics:

$$\dot{K}_i = I_i - \delta K_i, \quad K(0) = K_0, \quad i = 1, 2 \qquad (1.13)$$

with I as gross investment. The net revenue function $R_i(K_i)$ is written in a reduced form as

$$R_i(K_i) = K_i(a - K_1 - K_2)$$

He defines two types of strategies for the above differential game: an open-loop NE strategy pair $(\hat{I}_i, \hat{I}_j)$, where

$$\pi_i(\hat{I}_i, \hat{I}_j) \geq \pi_i(I_i, \hat{I}_j)$$

and a feedback strategy pair (I_i^*, I_j^*) where

$$\pi_i(I_i^*, I_j^*) \geq \pi_i(I_i, \hat{I}_j)$$

A stationary open-loop NE pair is one where the capacities of the two firms are constant, whereas the feedback strategy defines a set of decision rules that depend on time t and the current state. A firm using a feedback strategy does not precommit to a set of investment rates in advance but rather chooses its current investment rate based on current observed capacities.

An open-loop investment strategy by each firm is a best response to the path chosen by the rival. Two possible explanations may be given for precommitment in NE strategies. First, the firms are not able to observe the level of rival capacity after the start of the game. Secondly, open-loop NE investment strategies have a desirable dynamic consistency property; for example, if the game is truncated at t_1 with $t_1 < t$ and the open-loop NE strategies $(\hat{I}_1, \hat{I}_2)$ yielding a capacity pair $K_1(t_1), K_2(t_1)$ at $t_1 > 0$, then the pair $(\hat{I}_1, \hat{I}_2)$ is also an open-loop NE strategy pair for the truncated game beginning at t_1. Reynolds (1987) has shown that the dynamic system defined by Equation 1.11 yields a unique pair of open-loop NE investment strategies that yield stable capacity trajectories. These strategies are

$$\hat{I}_i = \delta K_i^i + (\lambda_1 + \delta)\left[\frac{(K_{i0} - K_{j0})}{2 - K_i^i}\right]\exp(\lambda_1 t) + (\lambda_2 + \delta)\left[\frac{(K_{i0} - K_{j0})}{2}\right]\exp(\lambda_2 t)$$

where

$$\lambda_1 = \frac{\left[r - \left(r^2 + 4\left((r+\delta)\,\delta + \frac{3}{c}\right)\right)^{1/2}\right]}{2} < 0$$

$$\lambda_2 = \frac{\left[r - \left(r^2 + 4\left((r+\delta)\delta + \frac{1}{c}\right)\right)^{1/2}\right]}{2} < 0$$

This follows from the Euler–Lagrange necessary condition

$$\ddot{K}_i - r\dot{K}_i - \left((r+\delta)\delta + \frac{2}{c}\right)\frac{K_i - K_j}{c} + \frac{(a - q(r+\delta))}{c} = 0$$

At a stationary point $\ddot{K}_i = \dot{K}_i = 0$ for $i = 1, 2$. This yields the stationary capacities at this equilibrium

$$K_i = \frac{(a - q(r+\delta))}{(3 + c\delta(R+\delta))} \tag{1.14}$$

Clearly as the adjustment cost rises, K_i falls.

Spence (1979) considered the case of irreversible investment where $I_i(t) \geq 0$. The incumbent here may use excess capacity as a device for entry deterrence. The incumbent being initially the sole supplier in the market can install capacity before a potential competitor makes an entry decision. Having installed a capacity in excess of the monopolist's output, the incumbent could threaten any potential entrant to increase output quickly as soon as entry occurs. This would reduce the entrant's post-entry profit and hence discourage entry. Hence in equilibrium the incumbent keeps the capacity which works as a potential entry barrier. It is the irreversibility of capital investment that allows the incumbent to commit himself to a certain output level before the entrant makes her entry decision. Using the Spence model, Schmalensee (1981) introduced some minimum level of output necessary for the entry process. This model shows that the case of entry deterrence is plausible only if economies of scale exist, that is due to first-mover advantage.

The Spence–Dixit model has been extended by Stehmann (1992), where the dominant firm adopts the leading role in a Stackelberg game, where it is assumed that the entrant believes that the established firm maintains its output in the face of potential entry. Two firms produce a homogeneous good, firm 1 being the incumbent and firm 2 the potential entrant. Both firms face the same technology, hence face the same cost function:

$$C_i(x) = \begin{matrix} cx_1 + rk_i + F, & \text{if } x_i \leq k_i \\ (c+r)x_i + F, & \text{if } x_i > k_i \end{matrix} \tag{1.15}$$

where x_i is output of firm i $(=1, 2)$, k_i is the capacity variable measured in the same unit as output and r is the rental cost of capital with F as fixed costs. In the first period the incumbent can choose his capacity. In the second both period firms decide on an output subject to a linear demand function

$$p = a - b(x_1 + x_2) \tag{1.16}$$

The first period capacity selection by firm 1 reduces its marginal costs below this capacity. The profit functions for the two firms are then

$$\pi_1 = \begin{matrix} [p(x_1 + x_2) - (c+r)]x_1 - F, & \text{if } x_1 > k_1 \\ [p(x_1 + x_2) - c]x_1 - rk_1 - F, & \text{if } x_1 \leq k_1 \end{matrix}$$

$$\pi_2 = [p(x_1 + x_2) - (c+r)]x_2 - F$$

Stehmann (1992) provides three conditions under which the incumbent may use excess capacity as a means of entry deterrence.

1. *Capacity condition* If the incumbent (firm 1) in the first period establishes the capacity necessary to deter and if firm 2 does not enter, then in the second period firm 1 as a monopolist has to produce below full capacity.
2. *Profitability condition* The second condition for excess capacity is that entry deterrence remains profitable for the incumbent. This depends on the structure of solutions of the post-entry game. Clearly in the post-entry game the incumbent will produce

$$x_1^L = \left(\frac{1}{2}\right) S_{cr},\ S_{cr} = \frac{\{a-(c+r)\}}{b}$$

while the entrant produces

$$x_2^f = \frac{S_{cr}}{4}$$

The respective profit is

$$\pi_1^L = \left(\frac{1}{8}\right) b\, S_{cr}^2 - F,\ \pi_2^f = \left(\frac{1}{16}\right) b\, S_{cr}^2 - F \qquad (1.17)$$

The firm 1 will prefer entry deterrence to accommodation if the monopolist's profit exceeds π_1^L as defined in Equation (1.17).
3. *Credibility condition* Also one has to investigate the conditions under which the threat to raise output to the entry-deterring capacity level is credible.

   We now determine the capacity level k_1^* which is necessary for firm 1 if entry has to be deterred. Clearly this capacity has to be such that for all positive x_2, $\pi_2(x_1 = k_1^*) \leq 0$. If $x_1 = k_1^*$, then the profit of firm 2 is

$$\pi_2 = [p(k_1^* + x_2) - (c+r)]x_2 - F$$

Maximizing π_2 yields the reaction function of firm 2 as

$$x_2 = \left(\frac{1}{2}\right)[S_{cr} - k_1^*]$$

This yields the profit function

$$\pi_2 = \left[bS_{cr} - b\left(\frac{S_{cr}}{2} - \frac{k_1^*}{2} + k_1^*\right)\right]\left(\left(\frac{1}{2}\right)(S_{cr} - k_1^*)\right) - F$$

On setting π_2 to zero, we obtain the critical level of capacity k_1^* which has to be installed to deter entry, where k_1^* is

$$k_1^* \geq S_{cr} - 2\left(\frac{F}{b}\right)^{1/2} \tag{1.18}$$

Three major conclusions are derived here. One is that the following three conditions have to obtain for excess capacity to prevail:

(i) $F < \left(\frac{b}{16}\right)\left(S_{cr} - \frac{r}{b}\right)^2$ capacity condition

(ii) $F > \left(\frac{b}{16}\right)\left[\frac{b}{r}\left(\frac{S_{cr}^2}{4} - \frac{1}{2}\left\{S_{cr} - \frac{r}{b}\right\}^2\right)\right]^2$ profitability condition

(iii) $F > \left(\frac{b}{16}\right)\left[S_{cr} - \frac{2r}{b}\right]^2$ credibility condition

The second result is that entry is blockaded if the unconstrained monopolist output already deters entry, i.e., $k_1^* < (1/2)S_{cr}$.

Thirdly, entry deterrence with excess capacity occurs when

$$\frac{S_c}{2} < k_1^* < \frac{1}{2}\left[S_{cr} + \frac{2r}{b}\right]$$

and the incumbent installs k_1^*. Here $S_c = (a - c)/b$.

Thus Stehmann has shown that three conditions have to be satisfied for excess capacity to hold in equilibrium, that is, the capacity condition which requires that the incumbent produces below capacity if entry is deterred, the profitability condition which requires that the entry deterrence offers the incumbent a higher profit than accommodation and the credibility condition which stipulates that the fixed costs F are above a certain level, that is

$$F > \left(\frac{b}{16}\right)\left[S_{cr} - \frac{2r}{b}\right]^2$$

so that the threat to produce $x_1 = k_1^*$ is credible.

1.4 Innovations and market dominance

In industry evolution a firm's growth is critically dependent on innovations which may be broadly conceived in Schumpeterian terms, which include R&D investment and other technical and organizational improvements. In this framework, innovations tend to provide several channels of potential market power which may deter future entry. Assume a Cournot type model where each firm *j* seeks to maximize profits

$$\pi_j = p(y_j + Y_{-j})y_j - C(y_j) \tag{1.19}$$

where y_j is the output of firm j, Y_{ij} is the total output of all its rivals, $C(y_j)$ is total cost of production with marginal cost $\hat{c}_j$. The optimal mark up μ can be easily derived as

$$\mu = \frac{(p - \hat{c}_j)}{p} = \frac{(1 + a_j)s_j}{e}$$

where $s_j = y_j/Y$ is the market share of firm *j*, $a_j = \partial Y_{-j}/\partial y_j$ and *e* is the price elasticity of demand. To arrive at the industry average mark up ($\bar{\mu}$) we weight the above equation by market shares and derive

$$\bar{\mu} = \frac{(p - \bar{c})}{p} = (1 + a)H \tag{1.20}$$

where $\bar{c} = \sum_{j=1}^{r} s_j\hat{c}_j$ and *H* is the Herfindahl index measuring concentration and it is assumed that the symmetry condition holds, that is $a_j = a$. The Schumpeterian theory of innovations assumes a stream of innovations whereby a successful innovator produces a newly invented good or service and continues to dominate the market (sometimes under the protection of patent rights) until driven out by the next innovator. Thus the relative success of the new entrant in driving out (or reducing the market share of) the incumbent depends on his ability to reduce unit cost. Since the rate of change in market share s_j can be viewed proportional to the cost differential,

$$\frac{\dot{s}_j}{s_j} = \lambda(\bar{c} - c_j)$$

where λ is the speed of selection at which firm shares react to their efficiency differences. On assuming λ, a_j and *e* to be fixed, the markup equation reduces to

$$\dot{\mu} = \lambda_\mu = \left[\left(\frac{1}{e}\right)\left\{\lambda(1 + a_j)s_j(\bar{c} - c_j)\right\}\right] \tag{1.21}$$

Note that as an industry-specific parameter a high λ would characterize an industry with a strong competitive adjustment mechanism much like the Walrasian process, while a low λ would imply a monopolistically competitive adjustment. In the latter case excess profits are competed away more slowly due to the higher concentration ratio. Kessides (1990) has established empirically that for US manufacturing industries the following tendencies persist:

$$\frac{\partial\lambda}{\partial H} > 0, \quad \frac{\partial\lambda}{\partial g} > 0, \quad \frac{\partial\lambda}{\partial \text{MES}} > 0 \quad \text{and} \quad \frac{\partial\lambda}{\partial K} > 0 \tag{1.22}$$

where g is the growth rate of total industry demand, MES is a measure of minimum efficient scale of output and K represents total capital required for the MES level. The roles of H, G, MES and K in dynamic competition are considered central to the model of hypercompetition developed by D'Aveni (1994) and Sengupta (2004). Sengupta has discussed three types of efficiency in this context: technological, access and resource (stronghold) efficiency. Resource efficiency is emphasized by the contribution of K in Equation (1.22), technological efficiency by MES and the access efficiency by H and g. By building barriers around a stronghold, the firm can reap monopoly profits. Access to distribution channels and low-cost supply sources in the supply chain and dynamic economies of scale may provide major barriers to trade by which the successful innovator firms may sustain a stronghold.

Since increased market share may be viewed as a proxy variable for entry, it is clear from Equation (1.21) that potential entry could be increased by reducing unit costs c_j below the industry average level $\bar{c}$, i.e., by increasing efficiency. In this context Cabral and Riordan (1994) have considered a game-theoretical model which posed the question, once ahead, what does the leading firm have to do to stay ahead? Here it is assumed that a firm's unit cost $c(s)$ is a decreasing function of cumulative past sales s. It is assumed that each firm's strategy depends only on the state of the game. Clearly the market dominance of the leading firm would depend on two dynamic effects: a cost effect and a strategy effect. Thus a firm at the very bottom of its learning curve maintains a strategic advantage as long as its rival has a higher cost. The strategic or prize effect refers to the potential prize from winning the lagging firm. Thus if the lagging firm has a sufficiently larger prize, then the prize effect could dominate the cost effect. In equilibrium the price difference of two firms would be proportional to the cost difference, that is $p_2 - p_1 = a(c_2 - c_1)$ where a is a positive constant, its higher value indicating a larger market dominance by the leading firm.

1.5 Conditions of entry and industry concentration

The factors which prevent entry being free are called "barriers to entry" by Bain and the level of price below which no firm finds it profitable to enter is called the *limit price*. Modigliani (1958) considered the case of scale economies and assumed that the incumbent and the potential entrant firms have cost curves which are sharply discontinuous at minimum optimal scale such that any output lower than that is infinitely costly to produce. Hence the entry limiting output is

$$q_{\mathrm{L}} = q_{\mathrm{c}}\left(1 - \frac{1}{s}\right) \tag{1.23}$$

where q_{c} is competitive output defined by price equal to marginal cost for all firms and s is the size of the market, $s = q_{\mathrm{c}}/\bar{q}$, where $\bar{q}$ is the minimum optimal scale. Limit price p_{L} then is given by the inverse demand function $p_{\mathrm{L}} = f(q_{\mathrm{L}})$. Thus if $p = \partial - \beta q$, then

$$q_{\mathrm{L}} = \frac{(\alpha - p_{\mathrm{L}})}{\beta}, \quad q_{\mathrm{c}} = \frac{(\alpha - p_{\mathrm{c}})}{\beta}$$

and it follows from Equation (1.23) that

$$p_{\mathrm{L}} = (\alpha - p_{\mathrm{c}})\left(1 - \frac{1}{s}\right) = p_{\mathrm{c}}\left(1 + \frac{1}{\varepsilon s}\right) \tag{1.24}$$

where p_{c} is competitive price and ε is the absolute value of the industry elasticity of demand at output q_{c}. Hence the price–cost margin corresponding to limit price can be written as

$$\frac{(p_{\mathrm{L}} - p_{\mathrm{c}})}{p_{\mathrm{L}}} = \left(\frac{\Pi}{R}\right)_{\mathrm{L}} = (1 + \varepsilon s)^{-1}$$

implying

$$\frac{\partial\left(\frac{\Pi}{R}\right)_{\mathrm{L}}}{\partial \varepsilon} < 0 \quad \text{and} \quad \frac{\partial\left(\frac{\Pi}{R}\right)_{\mathrm{L}}}{\partial s} < 0 \tag{1.25}$$

If we measure the importance of the economies of scale by $(1/s)$ then it follows that $\partial(\Pi/R)_{\mathrm{L}}/\partial\hat{s} > 0$, where $\hat{s} = 1/s$, that is the oligopolist's extra profit margin tends to increase with the importance of economies of scale and decrease with the industry price elasticity of demand which measures the market size. Note, however, that the economies of scale cannot be completely characterized by a minimum optimal scale; we have to include such features as absolute cost and product differentiation advantages. The Chamberlinian concept of "large group" under product differentiation, which defines the core framework of monopolistic competition may not yield any meaningful entry–exit equilibrium. As Stigler (1949) has noted, the diversity of demand and overall cost conditions may lead to the fact that there may be monopoly profits throughout the large group at equilibrium. It is not even clear that the equilibrium in the "large group" is even attainable. Under these vague and uncertain conditions disequilibria may prevail more often, that is market prices may continue to change and new firms may continue to enter and old firms continue to leave the group. Thus the concept of "industry" becomes completely devoid of any theoretical significance.

Thus the game-theoretic framework offers a more meaningful concept of entry–exit equilibrium than the monopolistic competition and we discuss a few theoretical models in this setup. The first model we consider in some detail is that of Sutton (1998) who has analyzed a two-stage game-theoretic model of NE in prices through the concept of "toughness of price competition". This concept of toughness refers to the functional relationship between market structure and profit margins per unit. The second model discusses a Stackelberg model, where there are economies of scale due to R&D spending and other learning curve effects. Here the potential entrant believes that the incumbent firm would maintain his output level and so acts as a Stackelberg follower, while the incumbent firm knows this to be the potential entrant's belief and so is able to act as a leader.

The Sutton model starts with N_0 firms indexed by $i = 1, 2, \ldots, N_0$. Firms invest in one or more R&D programs, each program having a particular trajectory m with its outcome indexed by a quality index μ_{im}. Firm i's capability is summarized by a *configuration* $u = (u_1, u_2, \ldots, u_i, \ldots, u_N)$, one for each of the N firms. We denote by u_{-i} all other firms excluding i. Denote by $u^* = \max_i \max_m u_{im}$ the highest level of competence attained along any trajectory. Now we introduce the process of entry and ask about the profit (expected or potential) of a new firm that enters on a single trajectory with a quality level k times greater than the maximum value u^* offered by any established firm. More specifically we ask, what is the minimum ratio of this profit to current industry sales that will be attained independently of the current configuration and the size of the market? For each k we define this ratio as

$$a(k) = \inf_u \left(\frac{\pi(ku^*|u)}{y(u)} \right) \tag{1.26}$$

where $\pi(\cdot)$ is profits and $y(u)$ is sales revenue. Then the following theorem is proved.

THEOREM 1 *(equilibrium entry). Given any pair $(k, a(k))$, a necessary condition for any configuration to be an equilibrium is that a firm offering the highest level of quality has a share of industry sales revenue exceeding $a(k)/k^\beta$, where the cost function is of the form $F(u) = F_0 u^\beta = \sum_{i=1}^{M} F_0 u_{im}^\beta$.*

PROOF. Consider any equilibrium configuration u in which the highest quality offered is u^* with its associated sales revenue y^*. Choose any firm belonging to this equilibrium configuration and denote its sales revenue by Sy^*, where the share of its industry sales revenue is $Sy^*/Sy(u)$ with S being the size of the total market. Now the definition of $a(k)$ implies that for a suitable trajectory the entrant's net profit is at least

$$aSy(u) - F(ku^*) = aSy(u) - k^\beta F(u^*)$$

where $a = a(k)$.

For the equilibrium the model requires the satisfaction of two conditions as follows:

1. *Stability condition* The entrant's net profit is not positive, that is

$$F(u^*) \geq \left(\frac{a}{k^\beta}\right) Sy(u) \tag{1.27}$$

2. *Viability condition* Each firm's final stage profit must cover its fixed outlays. Hence the sales revenue of the firm that offers quality u^* in equilibrium must equal or exceed its fixed outlays, that is

$$Sy^* > F(u^*) \geq \left(\frac{a}{k^\beta}\right) Sy(u)$$

   where its market share would be

$$\left(\frac{Sy^*}{(Sy(u)}\right) \geq \frac{a}{k^\beta} \tag{1.28}$$

On combining (1.27) and (1.28) we get the result mentioned in the theorem. □

Two implications of this result are most important for the entry condition. One arises when the industry consists of a large number of firms and the other when the "toughness of price competition" increases in the industry. In the first case the viability condition implies that each firm's spending on R&D is small relative to the industry's sales revenue. In this framework the returns to a high-spending entrant may be very large so that the stability condition is violated. Hence this framework cannot be an equilibrium configuration, that is if the concentration ratio C is very low, it cannot provide an equilibrium configuration. On using this result Sutton defines α as the highest value of the ratio:

$$\alpha = \sup_k \frac{a(k)}{k^\beta} \tag{1.29}$$

which is based on the viability condition shown in Equation (1.28). Here α serves as a measure of the extent to which a fragmented industry can be destabilized by the actions of a firm that outspends its many small rivals on R&D. This α measure is used to define a homogeneity index h as

$$h = \max_m \left(\frac{y_m(u)}{y(u)} \right)$$

where h represents the share of industry sales revenue accounted for by the largest product category. If all products have the same trajectory then $h = 1$. But h is close to zero if there are many different trajectories, each associated with a small group. By using this measure h and an equilibrium configuration one can derive the result

$$C_1 \geq h \left(\frac{a_0}{k_0^\beta} \right)$$

where C_1 is one-firm sales concentration ratio and the subscript zero denotes an equilibrium configuration. High alpha industries generally have both high R&D intensity and a high-level concentration. The opposite holds for low alpha industries. Thus one could use this result to empirically test the effect of concentration ratio on R&D intensity. For example we may refer to the empirical studies by Nooteboom and Vossen (1995) based on a survey of Dutch and US industries. They tested two questions: (1) Are large firms more innovative than small firms? and (2) Is innovation greater in more highly concentrated industries? In order to reduce heterogeneity of empirical data the industry effects of R&D investments and innovations are classified by four groups based on Pavitt's (1984) classification scheme as follows.

(1) Supplier-dominated industries like textiles, paper and fibers
(2) scale-intensive industries like rubber and plastics, metals, means of transport
(3) specialized industries like machinery, instrument and optical glass
(4) science- and technology-based industries such as chemicals and electrical goods.

The following specification is used for the regression model

$$\ln \text{R\&D} = a_0 + a_1 \ln S + a_2 \ln n + a_3 \ln C_4 \tag{1.30}$$

where R&D denotes annual expenditure on research and development, S is firm size, $n =$ number of firms conducting R&D and C_4 is the concentration ratio for four largest firms in the industry. The regression estimates of Equation (1.30) are as follows.

Table 1.1 Estimates of equation (1.30)

Category of firm	Dutch data (1989)				US data (1975)			
	Sample size	a_0	a_1	a_3	*Sample size*	a_0	a_1	a_3
Supplier-dominated	267	−4.88	0.533 (0.083)	–	86	−3.00	0.579 (0.062)	–
Scale-intensive	555	−5.43	0.905 (0.042)	–	231	−3.13	0.629 (0.038)	0.112 (0.043)
Specialized suppliers	288	−5.14	0.848 (0.072)	0.551 (0.119)	105	−2.49	0.583 (0.063)	–
Science based	182	−5.69	0.988 (0.045)	–	189	−3.87	0.883 (0.063)	–

The standard errors are in parentheses.

For these two different data sets one finds that for all Pavitt sectors except the science-based industries the R&D expenditure for firms increased less than proportionately with firm size. This implies that smaller firms are more R&D efficient. This effect is strongest in the supplier-dominated industries. For science-based industries there is no significant difference in this effect between small and large firms. Also we find a consistent positive effect of industry concentration, though it is not statistically significant in all cases. The authors conduct tests which show that the results of this empirical study are remarkably robust under change of data base and the introduction of additional explanatory variables.

Now we consider the problem of toughness of price competition, which may oftentimes violate the conditions of stability of NE. Instead of Cournot reaction curves in outputs we now consider the Bertrand mode with price reaction curves, which may characterize a Bertrand–Nash equilibrium. The toughness may then be measured by the variance of price fluctuations. Consider two suppliers ($i = 1, 2$) with demand and profit functions:

$$y_i = a_i - b_i p_i + c_i p_j - \hat{b}_i \dot{p}_i + \hat{c}_i \dot{p}_j$$
$$\pi_I = p_i y_i - C(y_i) \tag{1.31}$$

where $C(y_i)$ is the cost and the j-th firm ($j \neq i$) is the rival. The rate of change in price ($\dot{p}_i$) is due to investment in fixed capital like R&D investment in knowledge capital, for example

$$\dot{p}_i = \frac{\partial C}{\partial y_i} \frac{\partial y_i}{\partial K_i} \dot{K}_i$$

The objective function of firm i is

$$\max \int_0^\infty e^{-\rho t} \pi_i \, dt \tag{1.32}$$

where ρ is a positive discount rate assumed given.

The Bertrand dynamic model is more appropriate than the Cournot model when capacity and output can be easily adjusted, that is output is a short-run decision with respect to prices. Thus the firms set prices first and then output levels. In the Bertrand model firms simultaneously set prices and receive demand based on those prices. Implicitly the model assumes that firms produce an output exactly equal to the quantity demanded, that is output is perfectly adjusted to the quantity demanded at the prices initially set by firms.

The optimal solution of the Bertrand model can be easily derived from the differential game defined by Equations (1.31) and (1.32):

$$\dot{p}_1 = -\gamma_2 p_1 + \beta_2 p_2 - \alpha_2$$
$$\dot{p}_2 = \beta_1 p_1 - \gamma_1 p_2 - \alpha_1 \tag{1.33}$$

where

$$\alpha_i = \frac{a_i}{\hat{c}_i}, \beta_i = \frac{(2b_i + \hat{b}_i \rho)}{\hat{c}_i}, \gamma_i = \frac{c_i}{\hat{c}_i}$$

The characteristic equation for the homogeneous part of (1.33) is given by

$$\begin{vmatrix} -\gamma_2 - \lambda & \beta_2 \\ \beta_1 & -\gamma_1 - \lambda \end{vmatrix} = 0$$

That is,

$$\lambda^2 + (\gamma_1 + \gamma_2)\lambda + (\gamma_1\gamma_2 - \beta_1\beta_2) = 0$$

The two roots are

$$\lambda_1, \lambda_2 = \left(\frac{1}{2}\right)\left\{-(\gamma_1\gamma_2) \pm \sqrt{(\gamma_1\gamma_2)^2 - 4(\gamma_1\gamma_2 - \beta_1\beta_2)}\right\} \tag{1.34}$$

That is,

$$\lambda_1 = -2(\gamma_1 + \theta), \quad \lambda_2 = -2(\gamma_2 - \theta)$$
$$D = (\gamma_1 - \gamma_2) + 2\theta, \quad \theta > 0 \text{ and } D = \{(\gamma_1\gamma_2)^2 - 4(\gamma_1\gamma_2 - \beta_1\beta_2)\}$$

Clearly the two roots are real and these define a saddle point if $\gamma_2 < \theta$, $\theta > 0$ when $\lambda_1 < 0$, $\lambda_2 > 0$. But if $\gamma_2 > \theta$, $\theta > 0$ then $\lambda_1 < 0$, $\lambda_2 < 0$ define a stable region around the Bertrand–Nash equilibrium $(\bar{p}_1\bar{p}_2)$ defined by $\dot{p}_1 = \dot{p}_2 = 0$.

$$\bar{p}_1 = (\beta_1\beta_2 - \gamma_1\gamma_2)^{-1}(\gamma_1\alpha_2 + \beta_2\alpha_1)$$
$$\bar{p}_2 = (\beta_1\beta_2 - \gamma_1\gamma_2)^{-1}(\gamma_2\alpha_1 + \beta_1\alpha_2) \tag{1.35}$$

For economic realism we need the condition $\beta_1\beta_2 > \gamma_1\gamma_2$ when $\bar{p}_1$, $\bar{p}_2$ are positive. Note that as $\hat{c}_2$ rises, both α_2 and β_2 rise, implying a rise in $\bar{p}_1$. But if a_1 rises, implying a rise in α_1, then $\bar{p}_1$ also increases. A similar implication holds for $\bar{p}_2$.

Two comments may be added here. First, this model has to assume product differentiation to some degree. Otherwise if the product is homogeneous, then when $p_1 > p_2 > 0$, the demand for firm 1 will be driven to zero. Likewise firm 2 will have zero demand when $p_2 > p_1 > 0$. Secondly, the toughness of the price competition may be represented here by the speed of convergence to the Bertrand–Nash equilibrium point. Alternatively one could write the demand function for each i as

$$y_i = a_i - b_i p_i + c_i p_j - \hat{b}_i \dot{p}_i + \hat{c}_i \dot{p}_j + e_i$$

where e_i is the stationary error term assumed to be independently distributed with zero mean and variance σ_i^2. In this case we define a risk-adjusted profit function as

$$\tilde{\pi}_i = E(p_i y_i) - w_i p_i^2 \sigma_i^2$$

and use that in the objective function. One may assume for simplicity equal weights as $w_1 = w_2 = 0.5$. In this case the coefficient b_i in Equation (1.31) is changed to $(b_i + 0.5\sigma_i^2)$. Thus if σ_2^2 rises it may tend to raise β_2 in Equation (1.35) which implies a rise in the equilibrium price $\bar{p}_1$ of firm one. Also it may imply an increase in θ and if this increase is large so that γ_2 becomes less than θ, then it may impart a saddle point behavior, that is one trajectory moving away from the equilibrium and the other converging towards it.

1.6 Concluding remarks

The market entry game is intended to explain the two processes at work. One is the set of actions of firms intended to affect the current conduct of rivals and the other is altering the market structures in such ways that constrain the rival's subsequent strategies in future. Of several types of investment that the incumbent firms can make, limit pricing has been one of the major strategies which are invoked so as to discipline pricing decisions in the short run. The simplest limit pricing model which advances the proposition that potential competition disciplines short-run pricing decisions assumes homogeneous goods and scale economies and uses price alone as a strategic weapon and invokes the *Sylos postulate,* that is the entrant believes that incumbent will maintain the same output after entry to establish the entrant's conjecture about the post-entry equilibrium. An alternative view allows existing firms to tolerate entry if that is in their best interests and thus puts a lower bound on the limit price which one is likely to observe. Encaoua *et al.* (1986) have emphasized that the Sylos postulate is not a credible belief for large-scale entrants. This means one has to consider other weapons which can be used to discourage entry. For example advertising, investment in learning and R&D strategies provide other weapons. For example, an R&D strategy could create entry difficulties if a competitor's costs of product development lagged behind those of an early mover. The game-theoretic models we have surveyed here provide interesting insights into the question: What is the optimal mix of possible strategies or weapons that the incumbents may adopt in order to deter or reduce the intensity of the entry process? Questions of precommitment and credibility of any strategy mix are more important in dynamic game situations over time.

References

Bain, J.S. (1956) *Barriers to New Competition* (Cambridge, MA: Harvard University Press).

Cabral, I. and M. Riordan (1994) The learning curve, market dominance and predatory pricing, *Econometrica* 62, 1115–1140.

D'Aveni, R.A. (1994) *Hypercompetition: Managing the Dynamics of Strategic Maneuvering* (New York: Free Press).

Dixit, A. (1980) The role of investment in entry deterrence, *Economic Journal* 90, 95–106.

Encaoua, D., P. Geroski and A. Jacquemin (1986) Strategic competition and the persistence of dominant firms: A survey, in J. Stiglitz, and G. Mathewson (eds) *New Developments in the Analysis of Market Structure* (Cambridge: MIT Press).

Folster, S. and Trofimov, G. (1997) Industry evolution and R&d externalities. *Journal of Economic Dynamics and Control* 21, 1727–1746.

Fudenberg, D. and J. Tirole (1983) Capital as commitment: strategic investment to deter mobility, *Journal of Economic Theory* 32, 38–49.

Kessides, I.N. (1990) The persistence of profits in US manufacturing, in D.C. Mueller (ed.) *The Dynamics of Company Profits* (Cambridge: Cambridge University Press).

Modigliani, F. (1958) New developments on the oligopoly front, *Journal of Political Economy* 66, 213–232.

Nooteboom, B. and R. Vossen (1995) Firm size and efficiency in R&D spending, in A. Witteloostuijn (ed.) *Market Evolution: Competition and Cooperation* (Boston: Kluwer Academic Publisher).

Pavitt, S. (1984) Technology and market growth, *Review of Economic Studies* 51, 470–479.

Reynolds, S.S. (1987) Capacity investment, preemption and commitment in an infinite horizon model, *International Economic Review* 28, 69–88.

Schmalensee, R. (1981) Economies of scale and barriers to entry, *Journal of Political Economy* 89, 1228–1238.

Sengupta, J.K. (2004) *Competition and Growth: Innovations and Selection in Industry Evolution* (New York: Palgrave Macmillan).

Spence, A.M. (1979) Investment strategy and growth in a new market, *Bell Journal of Economics* 10, 1–19.

Stehmann, O. (1992) Entry deterrence and excess capacity in a Stackelberg game, *Economic Notes* 21, 450–467.

Stigler, G.J. (1949) The economics of scale, *Journal of Law and Economics* 1, 54–71.

Sutton, J. (1998) Technology and market structure. Cambridge, MIT Press.

Veloce, W. and A. Zellner (1985) Entry and empirical demand and supply analysis, *Journal of Econometrics* 30, 459–471.

2
Innovation and Efficiency in Industry Evolution

2.1 Introduction

In modern times the evolution of high-tech industries has been profoundly affected by innovations in different forms, for example new product designs, new software applications and new organizational forms. Schumpeter stressed two key elements as sources of evolutionary growth: new innovations in technology through the process of creative destruction and the access efficiency which involves racing up the escalation ladder in the stronghold arena. By building barriers under a stronghold, the firms reap monopoly profits as a first mover, and these profits can be used to fund aggressive price strategies and R&D investments.

We develop here a model where R&D investment represents dynamic innovations which affect the industry evolution, and it is empirically applied to modern high-tech industries: computers and pharmaceuticals. In both these cases the innovations stream in the form of R&D inputs has a profound impact on the composition of firms changing over time.

The R&D investments have several dynamic features of the innovations process which affect the industry evolution process. These features have significant economic implications for the selection and adjustment processes underlying the competitive and oligopolistic market structures. Our objective here is twofold: to discuss the different implications of R&D investment as it affects the competitive process of entry and exit in the evolution of industries and to relate innovations to the competitive edge of core competence by the leading firms. We discuss some theoretical models which incorporate spillovers of cooperative and noncooperative R&D expenditures and their impact on employment and growth and industry evolution. Since R&D expenditure usually involves increasing returns to scale (IRS), it directly affects the costs of entry and exit. We consider here a dynamic model of entry where the evolution of the number of firms in terms of entry and exit are endogenously determined, that is, the change in the number of firms is assumed to adjust up to the point where the costs of entry (of new firms) or exit (of existing firms) equal the net present value of entry. This type of model is closely related to the dynamic limit pricing model, where new firms are assumed to enter the industry as soon as the market price tends to exceed the limit price. R&D expenditures by some firms may reduce the limit price through reduction of unit costs through internal or external economies of scale. We consider some theoretical extensions of these dynamic models of R&D investment as it affects the equilibrium dynamics involving different number of firms and different R&D investment levels, which may imply multiple equilibria due to the nonconvexity of the underlying cost functions.

2.2 R&D investment and innovation efficiency

Innovations take many forms but in a broad sense they involve developing new processes, new products and new organizational improvements. R&D investment plays an active role in innovations in new processes and in new products and services. Several dynamic features of R&D investment by firms are important for selection and industry evolution. First of all, R&D expenditure not only generates new knowledge and information about new technical processes and products, but also enhances the firm's ability to assimilate, exploit and improve existing information and hence existing "knowledge capital". Enhancing this ability to assimilate and improve exiting information affects the learning process within an industry that has cumulative impact on the industry evolution. For example, Cohen and Levinthal (1989) have argued that one of the main reasons firms invested in R&D in semiconductor industry is that it provides an in-house technical capability that could keep these firms on the leading edge of the latest technology and thereby facilitate the assimilation of new technology developed elsewhere.

A second aspect of R&D investment within a firm is its spillover effect within an industry. R&D yields externalities in the sense that knowledge acquired in one firm spills over to other firms and very often knowledge spread in this way finds new applications both locally and globally and thereby stimulates further innovative activity in other firms.

Finally, the possibility of implicit or explicit collaboration in R&D networking or joint ventures increases the incentive of firms to invest. This may encourage more industry R&D investment in equilibrium. In the absence of collaboration the competing firms may not invest enough, since innovations cannot be appropriated by the inventor; for example, his competitors will copy the invention and thus "free ride" without paying for it. Thus the basic reason for the success of joint R&D ventures is that externalities or spillovers are internalized thus eliminating free rides.

We consider first the empirical basis of R&D innovations in modern industries and then its implications for selection and industry evolution. Cohen and Levinthal (1989) have made an important contribution in this area by analyzing the two faces of R&D investment in terms of spillover and externality. One impact of R&D spillovers emphasized by Nelson, Arrow and others is that they diminish the firm's incentive to invest in R&D and the related production. The other impact discussed by Cohen and Levinthal emphasizes the point that spillovers may encourage equilibrium industry R&D investment, since the firm's R&D investment develops its ability to exploit knowledge from the environment, that is develops its "absorptive" capacity or learning by which a firm can acquire outside knowledge. Thus a significant benefit of a firm's R&D investment is its contribution to the intra-industry knowledge base and learning, by which externality and spillovers may help firms develop new products and/or new processes.

The model developed by Cohen and Levinthal (C&L) starts with the firm's stock of knowledge and denotes the addition to firm's stock of technological and scientific knowledge by z_i and assumes that z_i increases the firm's gross earnings π^i but at a diminishing rate. The relationship determining z_i is assumed to be of the form

$$z_i = M_i + \gamma_i\left(\theta\sum_{j\neq i} M_j + T\right), \quad 0 \leq \gamma_i \leq 1 \tag{2.1}$$

where M_i is the firm's R&D investment, γ_i is the fraction of intra-industry knowledge that the firm is able to exploit, θ is the degree of intra-industry spillover of research knowledge. M_j represents other firms' $(j \neq i)$ R&D investments which contribute to z_i and θ denotes the degree to which the research effort of one firm may spill over to a pool of knowledge potentially available to all other firms; for example, $\theta = 1$ implies that all the benefits of one firm's research accrue to the research pool potentially available to all other firms, whereas $\theta = 0$ implies that the research benefits are exclusively appropriated by the firm conducting the research.

It is assumed that $\gamma_i = \gamma_i(M_i, \beta)$ depends on both M_i (the firm's R&D) and β, where β is a composite variable reflecting the characteristics of outside knowledge, that is, its complexity, ease of transferability and its link with the existing industry-specific knowledge. Clearly the composite variable β will differ from one industry to another; for example, in pharmaceutical industry it may involve lot of experimentation, long gestation periods and the complexity of the marketing process for new drugs, whereas for the computer industry it may involve software experimentation and the ease of application in multiple situations. It is assumed that the composite variable β denoting "ease of learning" is such that a higher level indicates that the firm's ability to assimilate outside knowledge is more dependent on its own R&D expenditure. Thus it is assumed that increasing β increases the marginal effect of R&D on the firm's absorptive capacity but diminishes the level of absorptive capacity.

Cohen and Levinthal evaluate the effects of increasing the explanatory variables such as β, θ and T on the equilibrium value of firm's R&D investment denoted by M^*, where it is derived from maximizing π^i with respect to M_i as

$$R = \text{MC} = 1 \tag{2.2}$$

where MC is the marginal cost of R&D expenditure equal to one and R is marginal return given by

$$R = \pi^i_{z_i}\left[1 + \gamma_{M_i}\left(\theta \sum_{j\neq i} M_j + T\right)\right] + \theta \sum_{j\neq i} \gamma_j \pi^i_{z_j} \tag{2.3}$$

where the subscripts denote partial derivatives. On solving Equations (2.2) and (2.3) simultaneously one obtains the equilibrium value of each firm's R&D denoted by M^*.

The impacts of the explanatory variables β, θ and T on M^* are derived as

$$\text{sign}\left(\frac{\partial M^*}{\partial \beta}\right) = \text{sign}\left[\pi^i_{z_i}\left\{\gamma_{M\beta}\left(\theta(n-1)M + T\right) + \theta(n-1)\right\}\frac{\partial \gamma}{\partial \beta}\pi^i_{z_i}\right] \tag{2.4}$$

$$\text{sign}\left(\frac{\partial M^*}{\partial \beta}\right) = \text{sign}\left[\pi^i_{z_i}\gamma_M\ (n-1)M + (n-1)\gamma\pi^i_{z_j}\right] \tag{2.5}$$

$$\text{sign}\left(\frac{\partial M^*}{\partial T}\right) = \text{sign}\left[\gamma_M \pi^i_{z_i} + \pi^i_{z_i z_j} + (n-1)\pi^i_{z_i z_j}\gamma(1 + \gamma_M T)\right] \tag{2.6}$$

The first term on the right-hand side of Equation (2.4) shows that a higher β induces the firm with more incentives to conduct R&D, because its own R&D has become more critical to assimilating its rivals' spillovers $\theta(n-1)M$ and the extra-industry knowledge T. The second term shows a decline in rivals' absorptive capacity $(n-1)\gamma$ as β increases. As a result the rival competitors are less able to exploit the firm's spillover. Due to both these effects the payoffs to the firm's R&D increases and *ceteris paribus* more R&D investment is induced.

The effect of θ on M^* is ambiguous, due to two offsetting effects: the benefit to the firm of increasing its absorptive capacity denoted by the first term, and the loss associated with the diminished appropriability of rents denoted by the second term on the right-hand side of Equation (2.5). Note, however, that the desire to assimilate knowledge generated by other firms provides a positive incentive to invest in R&D as θ increases.

The relation (2.5) shows that with an endogenous absorptive capacity, the firm has a positive incentive to invest in R&D to exploit the pool of external knowledge. With $\gamma_M = 0$, that is zero endogenous absorptive capacity, the sign $(\partial M^*/\partial T)$ is negative, since a higher T merely substitutes for the firm's own R&D, that is $\pi^i_{z_i z_j} < 0$.

Cohen and Levinthal estimate by regression (OLS, GLS and Tobit) models the effects of the knowledge inputs and other industry characteristics on unit R&D expenditure (intensity) of business units. The sample data included R&D performing business units consisting of 1302 units representing 297 firms in 151 lines of business in the US manufacturing sector over the period 1975–1977. The empirical data were obtained from Line of Business Program of the Federal Trade Commission (FTC) and the survey data collected by Levin *et al.* (1987). A set of estimates of selected regression coefficients is reproduced in Table 2.1. Appropriability here is defined as follows: the respondents in Levin *et al.* (1987) survey were asked to rate on a 7-point scale the effectiveness of different methods used by firms to protect the competitive advantages of new products an new processes. For a line of business appropriability is then defined as the maximum score. Thus if appropriability increases the spillover level declines and hence R&D intensity increases. The new plant variable is used to reflect the relative maturity of an industry's technology, that is it measures the percentage of an industry's plant and equipment installed with the five years preceding 1977 as reported in FTC's dataset. Industry demand conditions are represented by the price and income elasticity measures.

The explanatory variables T and β are measured indirectly for the survey data. The level of extra-industry knowledge T is measured by

Table 2.1 Effects of knowledge and other explanatory variables on R&D intensity

	OLS	GLS	Tobit
1. Technological opportunity			
(a) Appropriability $(1-\theta)$	0.396*	0.360**	0.260
	(0.156)	(0.104)	(0.161)
(b) Usertech	0.387**	0.409**	0.510**
	(0.99)	(0.070)	(0.166)
(c) Univtech	0.346**	0.245**	0.321*
	(0.128)	(0.089)	(0.147)
(d) Govtech	0.252*	0.170*	0.200*
	(0.100)	(0.076)	(0.100)
2. Basic science research			
(a) Biology	0.176	0.042	0.159
	(0.096)	(0.057)	(0.116)
(b) Chemistry	0.195**	0.095	0.149
	(0.071)	(0.050)	(0.078)
(c) Physics	0.189	0.037	0.156
	(0.109)	(0.082)	(0.109)
3. Applied science research			
(a) Computer science	0.336**	0.157	0.446**
	(0.123)	(0.093)	(0.121)
(b) Materials science	−0.005	−0.028	0.231*
	(0.121)	(0.089)	(0.116)
4. New plant	0.055**	0.041**	0.042**
	(0.008)	(0.006)	(0.007)
5. Elasticity of			
(a) Price	−0.180**	−0.071	−0.147*
	(0.061)	(0.044)	(0.060)
(b) Income	1.062**	0.638**	1.145**
	(0.170)	(0.136)	(0.180)
R^2	0.278	–	–

Note: Only a selected set of regression coefficient estimates are given here with standard errors in parentheses. * and ** denote significant values of t tests at 5 and 1 percent respectively.

five sources of which three are reported in Table 2.1, for example downstream users of industry's products (usertech), government agencies and research laboratories (govtech) and university research (univtech). The proxy variables used for β in Table 2.1 represent cumulativeness and the targeted quality of knowledge which are all field-specific, hence research in basic and applied sciences is reported here; for example, the characteristic that distinguishes the basic from the applied sciences is the degree to which research results are targeted to the specific needs of firms, where basic science is less targeted than the applied. Hence the β value associated with basic science research is higher than that associated with applied science. As a result the coefficient values of the technological opportunity variables associated with the basic sciences should exceed those of the applied sciences. The estimates in Table 2.1 show that except computer science the coefficients are uniformly greater for the basic sciences. The exception of computer science may also be due to the rapid advance in software and process development, where the basic and applied knowledge are intermingled.

2.3 Application in computer industry

Next we consider an application in the computer industry, which has witnessed rapid technological changes in recent years in both hardware and software R&D. Recent empirical studies have found cost-reducing effects of rapid technological progress to be substantial in most technology-intensive industries of today such as microelectronics, telecommunication and computers. Two types of productivity growth are associated with such technological progress: the scale economies effect and the shift of the production and cost frontiers. Also there exist substantial improvements in the quality of inputs and outputs. The contribution of R&D expenditure has played a significant role here. This role involves learning in different forms that help improve productive efficiency of firms. One may classify learning into two broad types: one associated with technological and the other with human capital. There are three types of measures of learning and general use in the literature. One is the cumulative experience embodied in cumulative output. The second measure is cumulative experience embodied in strategic inputs such as R&D investments in Arrow's learning-by-doing models. Finally, the experience in "knowledge capital" available to a firm due to spillover from other firms may be embodied in the cost function through the research inputs.

Unlike the regression approach of Cohen and Levinthal we now develop and apply a nonparametric and semiparametric model of production and cost-efficiency involving R&D expenditure and its learning effects. These nonparametric models do not use any specific form of the cost or production function; they are based on the observed levels of inputs, outputs and their growth over time. Technological progress (regress) is measured in this framework by the proportional rate of growth (decline) of total factor productivity (TFP), where TFP is defined as the ratio of aggregate (weighted) output to aggregate (weighted) inputs. The nonparametric efficiency model is specified here in terms of a series of linear programming (LP) models, the unifying theme of which is a convex hull method of characterizing the production frontier without using any market prices (i.e., technical efficiency) and the cost frontier (i.e., allocative efficiency which utilizes the input prices).

Consider now a standard input-oriented nonparametric model, also known as a DEA (data envelopment analysis) model for testing the relative efficiency of a reference firm or decision-making unit $h(\text{DMU}_h)$ in a cluster of N units, where each DMU_j produces s outputs (y_{rj}) with two types of inputs: m physical inputs (x_{ij}) and n R&D inputs as knowledge capital (z_{wj}):

$$\text{Min } \theta+\phi, \text{ subject to } \sum_{j=1}^{N} x_j\,\lambda_j \leq \theta X_h; \quad \sum_{j=1}^{N} Z_j\,\lambda_j \leq \phi Z_h \tag{2.7}$$

$$\sum_{j=1}^{N} Y_j\,\lambda_j \geq Y_h; \quad \sum_{j} \lambda_j \geq 0; \quad j=1,2,\ldots,N$$

Here X_j, Z_j and Y_j are the observed input and output vectors for each DMU_j, where $j=1,2,\ldots,N$. Let $\lambda^*=(\lambda_j^*)$, θ^*, ϕ^* be the optimal solutions of model shown in Equation (2.7) with all slacks zero. Then the reference unit or firm h is said to be technically efficient if $\theta^*=1.0=\phi^*$. If, however, θ^* and ϕ^* are positive but less than unity, then it is not technically efficient at the 100 percent level, since it uses excess inputs measured by $(1-\theta^*)\,x_{ih}$ and $(1-\phi^*)\,z_{wh}$. Overall efficiency (OE_j) of a unit j, however, combines with technical (TE_j) or production efficiency and the allocative (AE_j) or price efficiency as follows: $\text{OE}_j=\text{TE}_j\times\text{AE}_j$. To measure overall efficiency of a DMU_h one solves the cost-minimizing model:

$$\text{Min } C=c'x+q'z$$

$$\text{subject to} \quad X\lambda\leq x; \quad Z\lambda\leq z; \quad Y\lambda\geq Y_h; \quad \lambda'e=1; \quad \lambda\geq 0 \tag{2.8}$$

where e is a column vector with N elements each of which is unity, prime denotes transpose, c and q are unit cost vectors of the two types of inputs x and z which are now the decision variables and $X=(X_j)$, $Z=(Z_j)$ and $Y=(Y_j)$ are appropriate matrices of observed inputs and outputs. Denoting optimal values by *, technical efficiency is $\text{TE}_h=\theta^*+\phi^*$ as before, overall efficiency (OE_h) is C_h^*/C_h computed from model shown in Equation (2.8) and hence the allocative efficiency is $AE_h = C_h^*/(\theta^* + \phi^*)\ C_h$, where C_h and C_h^* are the observed and optimal costs for unit h respectively.

Now consider the special characteristics of the research inputs z. Since these inputs lower the initial unit production costs c_i and also affect the cost function nonlinearly, we can rewrite the objective function of Equation (2.2) as

$$\text{Min TC} = \sum_i \left[(c_i - f_i(\sum_w q_w z_w))x_i + \frac{1}{2} d_i x_i^2 \right] + \frac{1}{2} \sum_{w=1}^{n} g_w z_w^2 \quad (2.9)$$

subject to the constraints of the model shown in Equation (2.2). Here f_i is the unit cost reduction with $f_i < c_i$ and the component cost functions are assumed to be strictly convex, implying diminishing return to the underlying R&D production function. The optimal solutions z_w, x_i and λ_i now must satisfy the Kuhn–Tucker necessary conditions as follows:

$$\begin{aligned} f_i q_w x_i + \gamma_w &\leq g_w z_w, \quad z_w \geq 0 \\ f_i(\Sigma q_w z_w) + \beta_i &\leq c_i + d_i x_i, \quad x_i \geq 0 \end{aligned} \quad (2.10)$$

If the unit (DMU_h) is efficient with positive input levels and zero slacks, then we must have equality $\partial L/\partial z_w = 0 = \partial L/\partial x_i$, where L is the Lagrangean function. Hence we can write the optimal values (z_w^*, x_i^*) as

$$\begin{aligned} z_w^* &= \frac{(f_i q_w x_i^* + \gamma_w^*)}{g_w}, \quad w = 1, 2, \ldots, n \\ x_i^* &= \frac{(g_w z_w^* - \gamma_w^*)}{(f_i q_w)}, \quad i = 1, 2, \ldots, m \end{aligned} \quad (2.11)$$

By duality the production frontier for unit $j = 1, 2, \ldots, N$ satisfies

$$\alpha^{*\prime} Y_j \leq \alpha_0^* + \beta^{*\prime} X_j + \gamma^{*\prime} Z_j; \ (\alpha^*, \beta^*, \gamma^*) \geq 0$$

where the equality holds if unit j is efficient and there is no degeneracy due to congestion costs. Clearly a negative (positive or zero) value of α_0^* implies increasing (diminishing constant) returns to scale.

Note that the generalized quadratic programming model (Equation (2.9)) has many flexible features. First of all, if the research inputs are viewed as cumulative stream of past investment as in Arrow model of learning by doing, then the cost function TC in Equation (2.9) may be viewed as a long-run cost function. Given the capital input z^* the reference firm solves for the optimal current inputs x_i^* through minimizing the short-run cost function $TC(x|z^*)$. Second, the learning effect parameter $f_i > 0$ shows that the efficiency estimates through DEA model (Equation (2.8)) would be biased if it ignores the learning parameters. Third, the complementarity (i.e., interdependence) of the two types of inputs is clearly brought out in the linear relation between x_i^* and z_w in Equation (2.11). For example, it shows that

$$\frac{\partial x_i^*}{\partial z_w^*} > 0, \quad \frac{\partial x_i^*}{\partial f_i} > 0, \quad \frac{\partial x_i^*}{\partial \beta_i^*} > 0$$

$$\text{and} \quad \frac{\partial z_w^*}{\partial x_i^*} > 0, \quad \frac{\partial z_w^*}{\partial f_i} > 0, \quad \frac{\partial z_w^*}{\partial \gamma_w^*} > 0$$

Finally, compared to a linear program the quadratic programming model (Equation (2.9)) permits more substitution among the inputs, thus making it possible for more units to be efficient.

One limitation of the long-run cost (2.3) minimization model above is that it ignores the time profile of output generated by cumulative investment experience. Let $z(t) = (z_w(t))$ be the vector of gross investments and $k(t) = \int_0^t z(s)ds$ be the cumulative value, where

$$\dot{k}_w(t) = z_w(t) - \delta_w k_w(t) \tag{2.12}$$

where δ_w is the fixed rate of depreciation.

In this case the transformed DEA model becomes dynamic as follows:

$$\text{Min} \int_0^\infty e^{-\rho t} \, [c'(t) \, x(t) + C(z(t))] \, dt$$

subject to Equation (2.12) and the constraints of model shown in Equation (2.8).

Here $C(z(t))$ is a scalar adjustment cost, which is generally assumed nonlinear in the theory of investment. This type of formulation has been analyzed by Sengupta (2003), which shows the stability and adaptivity aspects of convergence to the optimal path.

Another type of characterization of the research inputs and their productivity is in the current literature of new growth theory. Thus Lucas (1993) considered a growth process where each firm has a production function, where its output depends on its own labor and physical capital inputs as well as the total knowledge capital of the whole industry. The availability of industry's knowledge capital occurs through the spillover mechanism or diffusion of the underlying information process. The utilization of the industry's knowledge capital by each firm has been called by Jovanovic (1997) as the learning effect which is very significant in the modern software-based industries. To characterize this learning effect we introduce a composite input vector X_j^C for DMU$_j$ as the share of each DMU$_j$ out of the industry total supply of each input, for example $\sum_{j=1}^{N} X_{ij}^C = X_i^T$, where X_i^T is the total industry supply of input *i*. We can then formalize the input-oriented DEA model in two forms as before:

$$\text{Min } \theta + \phi$$

$$\text{subject to } \sum_{j=1}^{N} X_j\lambda_j \leq \theta X_h; \quad \sum_j X_j^C \lambda_j \leq \phi X_h^C$$

$$\sum_j Y_j\lambda_j \geq Y_h; \quad \lambda' e = 1, \ \lambda \geq 0$$

and

$$\text{Min } C = c'x + q'x^c$$

$$\text{subject to } X\lambda \leq x; \quad X^C\lambda \leq x^C; \quad Y\lambda \geq Y_h; \quad \lambda' e = 1; \quad \lambda \geq 0 \quad (2.13)$$

On using the Lagrange multipliers α, β, γ and α_0, the production frontier for DMU$_h$ may be easily derived from the dual problem as

$$\alpha' Y_h = \beta' X_h + \gamma' X_h^C + \alpha_0; \quad \alpha, \ \beta, \ \gamma \geq 0$$

Clearly the interdependence of the two inputs x and x^C can be easily introduced in this framework through nonlinear interaction terms in the objective function of Equation (2.13) or through the method used in Equation (2.9) before.

Thus the generalized DEA models incorporate three additional sources of relative efficiency not found in the conventional DEA models: (1) unit cost reduction due to the complementarity effect of R&D inputs; (2) the IRS due to learning by doing and finally (3) the spillover effect of knowledge capital in the industry as a whole.

Now we consider an empirical application characterizing the role of R&D expenditure. An empirical application to a set of 12 firms (companies) in the US computer industry over a 12-year period (1987–1998) is used here to illustrate the concept of dynamic efficiency. The selection of 12 companies is made from a larger set of 40 companies over a 16-year period. R&D input is used here as a proxy for knowledge capital. We have used Standard and Poor's Compustat Database (SIC codes 3570 and 3571) for the input and output data for the 12-year period 1987–1998. This selection is based on two considerations: (1) survival of firms through the whole period and (2) promising current profit records. The companies are Compaq, Datapoint, Dell, Sequent, Data General, Hewlett-Packard, Hitachi, Toshiba, Apple, Maxwell, Silicon Graphics and Sun Microsystems.

The single output variable (y) is net sales in dollars per year and the nine inputs in dollars per year are combined into three composite inputs: x_1 for R&D expenditure, x_2 for net plant and equipment and x_3 for total manufacturing and marketing costs.

We consider input-based level and growth efficiency in terms of nonradial measures of efficiency, that is efficiency specific to each inputs. The level efficiency model is

$$\text{Min} \sum_{i=1}^{3} \theta_i$$

$$\text{subject to} \quad \sum_{j=1}^{12} y_j \lambda_j \geq y_k; \quad \sum_{j=1}^{12} x_{ij} \lambda_j \leq \theta_i x_{ik}, \quad i = 1, 2, 3 \tag{2.14}$$

$$\Sigma \lambda_j = 1, \quad \lambda_j \geq 0; \quad 0 \leq \theta_i \leq 1$$

and it is applied for three years 1987, 1991 and 1998. The growth efficiency model is used in the form

$$\text{Min} \quad \sum_{i=1}^{3} \phi_i$$

$$\text{subject to} \quad \sum_{j=1}^{12} \left(\frac{\Delta y_j}{y_j} \right) \mu_j \geq \frac{\Delta y_k}{y_k} \tag{2.15}$$

$$\sum_{j} \left(\frac{\Delta x_{ij}}{x_{ij}} \right) \mu_j \leq \phi_i \left(\frac{\Delta x_{ik}}{x_{ik}} \right), \quad i = 1, 2, 3$$

$$\Sigma \mu_j = 1, \quad \mu_j \geq 0; \quad 0 \leq \phi_i \leq 1$$

over the three-year periods 1987–1990, 1991–1994 and 1995–1998. Tables 2.2 and 2.3 present the empirical results. It is interesting to observe that a firm which is sequentially level efficient over the three years 1987, 1991 and 1998 is not necessarily growth efficient. Such an example is Hitachi, whose net sales did not grow as fast as some of its competitors.

Table 2.2 Nonradial measures (θ_i^*) of level efficiency

	1987			1991			1998		
	R&D (x_1)	Net plant (x_2)	Total production cost (x_3)	x_1	x_2	x_3	x_1	x_2	x_3
Data General	0.55	0.15	1.0	1.0	1.0	1.0	0.59	0.58	1.0
HP	1.0	1.0	1.0	1.0	1.0	1.0	1.0	1.0	1.0
Hitachi	1.0	1.0	1.0	1.0	1.0	1.0	1.0	1.0	1.0
Toshiba	1.0	1.0	1.0	1.0	0.96	0.97	1.0	1.0	1.0
Apple	1.0	1.0	1.0	1.0	1.0	1.0	0.40	0.69	1.0
Compaq	1.0	1.0	1.0	1.0	1.0	1.0	1.0	1.0	1.0
Datapoint	0.62	0.20	0.97	1.0	1.0	1.0	1.0	1.0	1.0
Dell	1.0	1.0	1.0	1.0	1.0	0.94	1.0	1.0	1.0
Maxwell	1.0	1.0	1.0	1.0	1.0	1.0	1.0	1.0	1.0
Sequent	1.0	1.0	1.0	1.0	1.0	1.0	0.64	0.53	1.0
Silicon Graphics	0.38	0.52	0.70	1.0	1.0	1.0	0.51	0.71	1.0
Sun Microsystems	0.57	0.29	1.0	1.0	1.0	1.0	1.0	1.0	1.0

Table 2.3 Nonradial measures (θ_i^*) of growth efficiency

	1987–1990			1991–1994			1995–1998		
	x_1	x_2	x_3	x_1	x_2	x_3	x_1	x_2	x_3
Data General	1.0	1.0	1.0	1.0	1.0	1.0	0.48	0.54	0.77
HP	0.49	1.0	0.47	0.55	0.80	1.0	1.0	1.0	1.0
Hitachi	0.40	0.68	0.65	1.0	1.0	1.0	0.94	0.84	1.0
Toshiba	0.29	0.62	0.72	1.0	1.0	1.0	1.0	1.0	1.0
Apple	0.52	0.69	0.64	0.51	0.44	0.76	1.0	1.0	1.0
Compaq	0.40	0.54	0.50	1.0	1.0	1.0	0.33	0.60	0.75
Data Point	1.0	1.0	1.0	1.0	1.0	1.0	1.0	1.0	1.0
Dell	0.61	0.44	0.47	1.0	1.0	1.0	1.0	1.0	1.0
Maxwell	1.0	1.0	1.0	1.0	1.0	1.0	1.0	1.0	1.0
Sequent	1.0	1.0	1.0	1.0	1.0	1.0	0.50	0.54	0.48
Silicon Graphics	1.0	1.0	1.0	1.0	1.0	1.0	0.25	0.25	0.38
Sun Microsystems	1.0	1.0	1.0	1.0	1.0	1.0	0.42	0.24	0.67

Measuring technological progress by the shadow price of the constraint $\Sigma\mu_j = 1$ the ranking of the 12 firms appears as follows:

1987–1990		1991–1994		1995–1998	
	Rank		Rank		Rank
Datapoint	1	Dell	1	Datapoint	1
Data General	2	Silicon	2	Maxwell	2
Sequent	3			Compaq	3
HP	4			Dell	4
Toshiba	5			Data General	5
Sun	6			Sequent	6
Hitachi	7			Silicon	7
Apple	8				
Compaq	9				
Silicon	10				

Note that during 1991–1994 only Dell and Silicon Graphics exhibited positive technological progress, others with zero or negative technological progress. Datapoint regained first rank during the period 1995–1998. Table 2.2 also shows Datapoint to be growth efficient in terms of all the three inputs over all the periods. On writing the dual formulation of the LP model (2.15) the dynamic production frontier for the efficient firm k may be written as

$$\alpha^* \left(\frac{\Delta y_k}{y_k}\right) = \beta_0^* + \sum_{i=1}^{3} \beta_i^* \left(\frac{\Delta x_{ik}}{x_{ik}}\right) \tag{2.16}$$

The values of the technological progress ($\beta_0^* > 0$) or regress ($\beta_0^* < 0$) appear as follows:

	1987–1990		1995–1998
Datapoint	10.74	Datapoint	19.35
Data General	3.32	Maxwell	7.36
Sequent	3.24	Compaq	1.86
Toshiba	0.261	Sequent	0.38
Sun	0.260	Silicon	0.10
Hitachi	0.25		
Apple	0.08		
Compaq	0.040		
Silicon	0.041		

Measuring Solow-type technological progress by the ratio β_0^*/α^* we obtain the following results:

(i) Technological progress is above 4 percent per year for 58 percent of firms in 1987–1990, 83 percent in 1991–1994 and 75 percent in 1995–1998.
(ii) Some typical examples are (in percents)

1987–1990 HPC (6.4), DGC (4.9), DEL (3.8) and SUN (4.9)
1991–1994 DGC (5.1), HPC (9.0), APL (9.1) and SIL (6.4)
1995–1998 DGC (6.1), HIT (9.0), COM (16.4) and MAX (7.7)

The sum of the input coefficients $\sum_{i=1}^{3} \beta_i$ measuring returns to scale is equal to or exceeds unity in all cases of the sample of efficient firms.

A second way of analyzing the empirical results is to run a regression of the dependent variable log output $= \hat{y}$ on the three independent variables: log R&D ($\hat{x}_1$), log plant and equipment ($\hat{x}_2$) and log cost of goods sold ($\hat{x}_3$) with a dummy variable D for each coefficient, where

$D = 1.0$ for the efficient firms and zero otherwise. The results are as follows:

$$
\begin{aligned}
\text{1987–1998} \quad & \hat{y} = 1.199^{**} + 0.162^{**}\,\hat{x}_1 + 0.065^{*}\,D\hat{x}_1 \\
& + 0.009\,\hat{x}_2 - 0.034\,D\hat{x}_2 \\
& + 0.743^{**}\,\hat{x}_3 + 0.034^{*}\,D\hat{x}_3 \quad (R^2 = 0.996) \\
\text{1991} \quad & \hat{y} = 1.214^{**} + 0.262^{**}\,\hat{x}_1 - 0.075\,\hat{x}_2 \\
& + 0.791^{**}\,\hat{x}_3 \quad (R^2 = 0.998) \\
& \text{D significant for } \hat{x}_1 \text{ and } \hat{x}_3 \text{ only} \\
\text{1998} \quad & \hat{y} = 0.925^{**} + 0.140^{*}\,\hat{x}_1 + 0.015\,\hat{x}_2 + 0.0842^{**}\,\hat{x}_3 \\
& (R^2 = 0.998) \quad \text{D significant for } \hat{x}_1 \text{ and } \hat{x}_3 \text{ only}
\end{aligned}
$$

Clearly R*&D expenditures have played a most dynamic role in the productivity growth of the efficient firms in the computer industry and this trend is likely to continue in the future.

To summarize we note that the role of knowledge capital and its cost-reducing effects are incorporated here in order to capture learning by doing. The recent upsurge in the personal computer industry is illustrated here in terms of the two types of efficiency models: one deals with the level efficiency and the other with the growth efficiency. Solow-type technological progress is estimated by the growth efficiency model applied to the computer industry over the period 1987–1998. Two important findings are that all the growth efficient firms exhibited technological progress of or above 4 percent per year on the average and the R&D input is significantly more important for the efficient firms than the inefficient ones.

Now we consider the impact of own R&D expenditure and compare it with the spillover effect for the companies in the computer industry. Denoting proportional changes by a hat, let $\hat{c}_j$, $\hat{y}_j$, $\hat{x}_{1j}$ and $\hat{x}_{2j}$ indicate changes in average costs, output, own R&D expenditure and the aggregate industry R&D spending available for firm j, then the dynamic efficiency model is used in the form

$$
\begin{aligned}
& \text{Min } \theta \\
& \text{subject to } \sum_{j=1}^{N} \hat{c}_j \lambda_j \le \theta \hat{c}_h; \quad \sum_{j=1}^{N} \hat{y}_j \lambda_j \ge \hat{y}_h
\end{aligned}
$$

$$\sum_{j=1}^{N} \hat{x}_{1j}\lambda_j \le \hat{x}_{ih}; \quad \sum_{j=1}^{N} \hat{x}_{2j}\lambda_j \le \hat{x}_{2h} \tag{2.17}$$

$$\sum_{j=1}^{N} \lambda_j = 1, \lambda_j \ge 0; \quad j = 1, 2, \ldots, N$$

Here $x_{2j} = \sum_{\substack{k \neq j \\ k=1}}^{N} x_{ik}$ is used as a proxy for industry-wide R&D knowledge capital. This is analogous to the measure M_j used by Cohen and Levinthal indicating other firms' R&D expenditure.

This model (Equation (2.17)) is empirically applied to the computer industry data from Compustat over the period 1985–2000. The initial dataset for 40 companies was reduced to 12 because of heterogeneity and continuous availability consideration. Thus the analyzed data set consist mostly of hardware-based companies, since the constraint of homogeneity is considered important for picking companies in the portfolio. Perfect homogeneity would exist if each company and its products could be perfect substitutes for another company. However, our selected data set only represents broad homogeneity and not perfect homogeneity.

The detailed characteristics of the empirical dataset have been analyzed by Sengupta (2003, 2004). Using net sales as output, average cost excluding R&D expenditure is computed from total cost per unit of output for each company. Standard and Poor's Compustat data are used for calculating total cost as the sum of the following major cost components: R&D expenditure, net plant and equipment and total manufacturing and marketing costs as in Equation (2.22).

The 16-year period 1985–2000 is divided into three subperiods 1985–1990, 1990–1995, 1995–2000 and 1997–2000, in order to convey the trend in terms of moving averages. It is clear from the LP model (Equation (2.17)) that for a dynamically efficient firm j we would have θ close to unity, and the cost frontier would appear as

$$\frac{\Delta c_j}{c_j} = \beta_0 - \beta_1 \left(\frac{\Delta y_j}{y_j}\right) - \beta_2 \left(\frac{\Delta x_{1j}}{x_{1j}}\right) - \beta_3 \left(\frac{\Delta x_{2j}}{x_{2j}}\right)$$

$$\beta_1, \beta_2, \beta_2 \ge 0, \ \beta_0 \text{ free in sign} \tag{2.18}$$

where β_i's are appropriate optimal values of Lagrange multipliers. Here β_2 and β_3 measure the impact of own and outside R&D spending respectively.

The average estimated values of β_2 and β_3 for selected efficient companies are as follows:

	1985–1990		1990–1995		1995–2000	
	β_2	β_3	β_2	β_3	β_2	β_3
Apple	1.23	0.35	1.26	0.31	0.94	0.02
Dell	2.49	0.47	0.74	0.05	1.10	0.31
HP	1.80	0.38	1.03	0.65	0.94	0.22
Silicon Graphics	0.91	0.19	0.95	0.21	1.42	0.75
Sun	1.05	0.24	0.85	0.25	1.01	0.23

If we run Equation (2.18) as a regression model for 12 firms over the periods 1985–1992 and 1992–2000 the results appear as follows:

	β_0	β_1	β_2	β_3	R^2
1985–1992	−0.01*	0.004	1.25*	0.25	0.96
1992–2000	−0.02**	0.01**	1.41*	0.31*	0.92

Here * and ** denote significance of t values at 5 and 1 percent levels respectively. Two points emerge very clearly from these growth efficiency results. First, the impact of R&D growth is highly significant; this is particularly so for the leading firms. Since data are not separately available for hardware- and software-specific R&D expenditure, it is not possible to estimate their separate impact. But since average computer prices have followed a continuous downtrend, it may be reasonable to assume that this downtrend has been largely cost driven. Secondly, the spillover effect has consistently followed the cost-reducing tendency, although at a lesser rate than the overall R&D spending. Recent outsourcing of R&D spending abroad may also intensify this tendency of spillover effect to be more and more important.

2.4 Core competence and industry evolution

What makes a firm grow? What causes an industry to evolve and progress? From a broad standpoint two types of answers have been offered. One is managerial, the other economic. The managerial perspective is based on organization theory, which focuses on the cost competence as the primary source of growth. The economic perspective emphasizes productivity and efficiency as the basic source of growth.

Economic efficiency of both physical and human capital including innovations through R&D have been stressed by the modern theory of endogenous growth. Core competence rather than market power has been identified by Prahalad and Hamel (1994) as the basic cornerstone of success in the modern hypercompetitive world of today. Core competence has been defined as the collective learning of the organization, especially learning how to coordinate diverse production skills and integrate multiple streams of technologies. Four basic elements of core competence are as follows: learn from own and outside research, coordinate, integrate so as to reduce unit costs and innovate so as to gain market share through price and cost reductions.

A company's own R&D expenditures help reduce its long-run unit costs and also yield spillover externalities. These spillovers yield IRS as discussed before. Now we consider a dynamic model of industry evolution, where R&D investments tend to reduce unit costs and hence profitability. This profitability induces new entry and also increased market share by the incumbent firms who succeed in following the cost-efficiency frontier.

Denoting price and output by p and y respectively, the dynamic model may be specified as

$$\dot{y} = a(p - c(u)), \quad a > 0 \tag{2.19}$$

where dot denotes time derivative and $c(u)$ is the average cost depending on innovation u in the form of R&D expenditure. When total profit $\pi = (p - c(u))$, y is positive, it induces entry in the form of increased output over time. Entry can also be represented by $\dot{n}$, where n denotes the number of firms, but we use $\dot{y}$ since n is discrete. We assume that each incumbent firm chooses the time path $u(t)$ of R&D that maximizes the present value $v_0 = \int_0^\infty \exp(-rt)\pi(u)\,dt$ of future profits as the known discount rate r. The current value of profits at time t is

$$v(t) = \int_t^\infty \exp(-r(\tau - t))\pi(u)d\tau$$

On differentiating $v(t)$ one obtains

$$\dot{v} = rv - \pi(u) \tag{2.20}$$

This represents capital market efficiency or the absence of arbitrage.

The dynamic model defined by Equations (2.19) and (2.20) is a model of industry evolution. When excess profits is zero, one obtains the equilibrium $p = c(u^*) = c(y^*, u^*)$. Again if the cost of entry equals the net present value of entry $v(t)$, then $\pi(u^*) = rz$, z being the cost of entry with $z = v$. The dynamics of the evolution model can be discussed in terms of a linearized version of equations (2.19) and (2.20) and the associated characteristic roots. Sengupta (2004) has analyzed the stability aspects of this dynamic model elsewhere.

The profitability equation (2.19) may also be written in terms of the market share s of the incumbent firm as

$$\dot{s} = b(\bar{c} - c(u)) \tag{2.21}$$

when price is assumed to be proportional to the industry average cost function $\bar{c}$, where $\bar{c}$ is the average of both best practice firms and others. Mazzacato (2000) and Sengupta (2004) have applied this type of dynamics of market evolution in several industries. Whenever $\bar{c} > c(u)$ the incumbent firm increases its market share. Also by the optimal allocation of R&D innovations u, the incumbent firms may succeed in reducing unit costs $c(u)$ in the long run. This also increases their market shares. Following the Fisherian model of growth of fitness in natural evolution of species, Mazzacato (2000) has shown that the rate of change in industry average cost function may be viewed as proportional to the variance of individual costs $c_i(u_i)$ so that

$$\frac{d\bar{c}}{dt} = \dot{\bar{c}} = \alpha\sigma^2(t)$$

More generally it may be written as

$$\dot{\bar{c}} = \alpha_1\sigma^2(t) - \alpha_2\bar{c} \tag{2.22}$$

where α_1, α_2 are nonnegative coefficients. Thus if cost variances rise, the industry average cost rises, implying a fall in overall efficiency. On the other hand, if $\bar{c}$ rises, it tends to reduce the growth rate of $\bar{c}$ over time due to more exits. This impact of heterogeneity in costs has sometimes been called the "churning effect" by Lansbury and Mayes (1996), who analyzed the entry–exit dynamics of several industries in the United Kingdom.

We apply these dynamic evolution models empirically over the pharmaceutical industry, where the R&D inputs are very important.

2.5 Application in pharmaceutical industry

We consider in this section some empirical applications of the industry evolution models for the pharmaceutical industry over the period 1981–2000. The pharmaceutical industry is important in the R&D investment perspective since research and innovations are sources of potential competition between rival firms. Also it has significant impact on net sales, when a growth perspective is adopted. Growth of R&D affects demand growth through both product innovations and brand loyalty.

We have used the empirical data on costs and output (net sales) for the pharmaceutical industry over the period 1981–2000. A set of 17 companies out of a larger set of 45 is selected from the Compustat database available from Standard and Poor. This selection is based on considerations of continuous availability of data on R&D expense and its share of total costs. The selected companies comprise such well-known firms as Merck, Eli Lily, Pfizer, Bausch and Lomb, Johnson and Johnson, Glaxosmithkline, Schering-Plough and Genentech. The share of R&D in total costs is quite important for these companies.

The distribution of net sales over the period 1981–2000 for these 17 companies is as follows:

	1981	1990	2000
Mean	2061.07	4263.24	4263.24
Standard deviation	2125.07	383.69	518.22
Skewness	0.8317	0.2973	0.7458

Clearly the data are more homogenous in years 1990 and 2000 compared to the year 1981.

Four types of estimates are calculated for the selected companies in the pharmaceutical industry. Table 2.4 provides the estimates of cost-efficiency along the total cost frontier. The model here is of the form

$$\text{Min } \theta$$

$$\text{subject to } \sum_{j=1}^{n} C_j\lambda_j \leq \theta C_h; \quad \sum_{j=1}^{n} x_j\lambda_j \leq x_h$$

$$\Sigma y_j\lambda_j \geq y_h; \quad \Sigma\lambda_j = 1, \quad \lambda j \geq 0 \qquad (2.23)$$

$$j = 1, 2, \ldots, n$$

Table 2.4 Efficiency coefficients (θ^*) for the total cost and the average cost frontier

	1981		1990		2000	
	TC	AC	TC	AC	TC	AC
Abbott Lab	0.829	0.831	0.871	0.885	0.807	0.832
Alza Corp	0.312	0.324	0.452	0.453	0.800	0.802
American Home Products	1.000	1.000	1.000	1.000	0.771	0.809
Bausch & Lomb	0.877	1.000	0.768	1.000	0.737	0.739
Bristol Myers	0.832	0.861	0.971	1.000	0.939	0.982
Forest Lab	0.878	1.000	0.661	0.662	0.531	0.532
Genentech	0.264	0.273	0.549	0.559	0.545	0.556
Glaxosmith	0.493	0.514	0.787	0.818	0.847	0.964
IGI Inc	1.000	0.024	1.000	0.709	1.000	1.000
Johnson & Johnson	0.958	1.000	1.000	1.000	0.938	1.000
Eli Lily	0.772	0.886	0.781	0.811	0.840	0.903
Merck	0.710	0.848	0.983	1.000	1.000	1.000
MGI Pharma	0.548	0.196	1.000	0.199	0.6800	0.442
Natures Sunshine	0.432	1.000	1.000	1.000	1.000	1.000
Pfizer Inc	0.764	0.822	0.838	0.822	0.841	1.000
Schering-Plough	0.703	0.709	0.796	0.808	0.837	0.872

where firm h is the reference firm with output y_h and costs C_h and x_h. Here x_h is R&D costs and C_h is total costs excluding R and R costs. Total costs comprise cost of goods sold, net plant and machinery expenditure and all marketing costs excluding R&D expenses denoted by x_h. A growth efficiency form of this model is specified in Equation (2.17) before. The optimal values of the LP model denoted by * are such that

$$\theta^* = 1 \text{ with } \sum C_j \lambda_j^* = C_h; \quad \sum \lambda_j^* = x_h$$

then the firm h is efficient, that is it lies on the cost-efficiency frontier; also the R&D inputs are optimally used. If, however, $\theta^* < 1$, then $\sum C_j \lambda_j^* < C_h$, indicating that optimal costs $C_h^* = \sum C_j \lambda_j^*$ are lower than the observed costs C_h. Hence the firm is not on the cost-efficiency frontier.

A second type of estimate uses the growth efficiency model (Equation (2.17)) to characterize the efficient and inefficient firms and then applies the regression model in order to estimate the impact of the growth of R&D inputs. This is compared with the level effect, when we regress total cost on R&D and other variables. A third type of estimate calculates the impact of e and other component inputs on total sales revenue for firms which are on the cost-efficiency frontier.

Finally, we estimate the market share models (2.21) and (2.22 & 2.23) where cost-efficient firms are tested if their share has increased when the R&D inputs helped reduce their average costs.

Table 2.4 reports the optimal values θ^* of the LP model (2.22) for each firm for three selected years 1981, 1990 and 2000. If instead of total costs (TC), we use average costs (AC) defined by the ratio of total costs to net sales, the estimates of θ^* change but not very significantly. Table 2.5 presents a summary of firms which are efficient in terms of TC, AC and RD level.

Tables 2.6 and 2.7 report the estimates of the cost frontier in two forms: the level form and the growth form, where the inputs are separately used as an explanatory variable. Here the growth form exhibits much better results over the level form.

Finally Table 2.8 reports the estimates of the market share models, when each firm is analyzed over the whole period. The market share

Table 2.5 Number of efficient firms with efficient TC, AC and R&D

	TC	AC	R&D level
1981	3 (18%)	6 (35%)	6 (35%)
1990	5 (29%)	6 (35%)	6 (35%)
2000	3 (18%)	5 (29%)	5 (29%)

Table 2.6 Cost frontier estimates of selected firms over the whole period 1981–2000, $TC_j = a + by_j + C\hat{R}_j + d\theta_j$

Firm	*a*	*b*	*c*	*d*	R^2	*F* statistics
ABT	1355**	1.301**	−295.3*	−0.046[a]	0.999	3933.2
AHP	15410	0.645	−5.8E07	−13068.7	0.999	4402.3
BOL	−250.3	1.608**	N	−1375.4*	0.996	780.22
GSK	1915.1**	1.335**	−10929.1	−1505.7*	0.993	378.4
IG	−37.17	3.112**	245.1	−1.510[a]	0.987	221.1
PF	1672.7**	1.350**	−46652.7**	−0.064[a]	0.999	4196.6
PHA	−16807.7*	2.224**		−1.517*[a]	0.985	187.2

* and ** denote significance of *t* values at 5 and 1 percent respectively; the superscript a denotes the cross-product of efficiency and output levels as the repressor, since the output term was highly dominant. TC_j, y_j and θ_j are total cost, output and efficiency scores. $\hat{R}_j$ is a proxy for R&D combined with output; *N* denotes a high value which is not significant at even 20 percent level of *t* test; for other firms not included here multicollinearity yields singularity of estimates, hence these are not reported.

Table 2.7 Sources of growth of total costs for the industry as a whole, $\mathrm{GTC}_j = a + b\mathrm{GRD}_j + \mathrm{G}y_j$

	1982	1991	2000
a	−0.124	0.097**(*D*)	−0.165**
b	0.914**	0.389**(*D*)	0.815**
c	6.15E−06	−6.2E−06*	9.11E−06**
R^2	0.871	0.653	0.913
F	47.21	13.176	73.424

GTC, GRD and Gy denote the proportional growth rates of total costs, R&D and total output respectively; * and ** denote significant *t* values at 5 and 1 percent; *D* denotes a dummy variable with one for the efficient units and zero for others. It indicates that these coefficients are significantly different for the efficient firms compared to the inefficient ones.

Table 2.8 Estimates of market share models for selected firms in the pharmaceutical industry (1981–2000), $\Delta s = b_0 + b_1(\bar{c} - c(u))$

Firm	b_0	b_1	R^2	F
ABT	1.07**	2.267**	0.385	5.002
AHP	1.295**	4.299	0.155	1.464
BOL	1.095**	0.691[a]	0.296	3.156
BM	1.022**	1.577*	0.224	2.312
GSK	1.082**	0.727	0.027	0.221
PF	1.209**	12.271**[a]	0.492	7.750
PHA	1.121**	3.947*	0.198	1.980

* and ** denote significance of *t* values at 5 and 1 percent respectively; the superscript a denotes that the quadratic term $(\bar{c} - c(u))^2$ has a significant positive coefficient.

model predicts that the efficient firms would increase their market shares when the industry average cost rises due to the failure of inefficient firms to reduce their long-run average cost. For the whole industry over the period 1981–2000 this relationship is tested by the following regressions.

$$\bar{c} = \underset{(t=10.02)}{3953.2^{**}} + \underset{(8.323)}{5.79^{**}} - 05\sigma^2;\ R^2 = 0.912;\ F = 186.98$$

$$\Delta\bar{c} = \underset{(t=2.31)}{0.145^{**}} + \underset{(3.45)}{0.005^{**}\sigma^2} - \underset{(1.91)}{0.014^{*}}\ \bar{c};\ R^2 = 0.721;\ F = 291.01$$

Clearly the churning effect is found to be important for this industry.

Several points emerge from the estimated results in Tables 2.4–2.8. First of all, the number of firms on the cost-efficiency frontier is about one-third and these firms are invariably efficient in using their R&D inputs. Secondly, both the efficiency score and the composite R&D inputs help the firms improve their cost-efficiency and these results are statistically significant. Growth of R&D inputs is as important as output growth in contributing to the increase of costs over time. This implies that the R&D inputs play a very dominant role in the growth of the pharmaceutical industry. It also increases profit through higher demand. Thirdly, the market share model shows very clearly that the more efficient firms with $\bar{c} > c(u)$ increase their market share over time and the two sources of this share gain are the decrease in average cost through R&D and other forms of innovation and the increase in industry-wide average cost due to the failure of less efficient firms to reduce their long-run average costs. Clearly when the cost heterogeneity measured by cost variance σ^2 rises, it tends to increase the industry average cost ($\bar{c}$) over time. This creates a long-run force for increased entry and/or increased market share.

2.6 Concluding remarks

The role of efficiency improvement through innovations and R&D inputs in explaining the growth of firms is analyzed here through cost-efficiency models. These models are empirically applied to the computer and pharmaceutical industries. Broadly the results confirm that the innovations in the form of R&D inputs have played a significant role in improving efficiency for selected firms in these industries and these firms have succeeded in improving their average market share. Thus comparative efficiency lies at the core of industry evolution.

References

Cohen, W.M. and D.A. Levinthal (1989) Innovation and learning: The two facets of R&D, *Economic Journal* 99, 569–596.

Jovanovic, B. (1997) Learning and growth, in D.M. Kreps and K.F. Wallis (eds) *Advances in Economics and Econometrics* (Cambridge: Cambridge University Press).

Lansbury, M. and D. Mayes (1996) Entry, exit, ownership and the growth of productivity, in D. Mayes (ed.) *Sources of Productivity Growth* (Cambridge: Cambridge University Press).

Levin, R., R. Nelson and S.G. Winter (1987) Appropriating the returns for industrial R&D, *Brookings on Economic Activity*, Washington, DC.

Lucas, R.E. (1993) Making a miracle, *Econometrica* 61, 251–272.

Mazzacato, M. (2000) *Firm Size, Innovation and Market Structure* (Cheltenham: Edward Elgar).
Prahalad, C.K. and Hamel, G. (1994) Competing for the future. Cambridge, Harvard University Press.
Sengupta, J.K. (2003) *New Efficiency Theory with Applications of Data Envelopment Analysis* (Heidelberg: Springer).
Sengupta, J.K. (2004) *Competition and Growth: Innovations and Selection in Industry Evolution* (New York: Palgrave Macmillan).

3
The Costs and Effects of Market Entry

3.1 Introduction

Evolution of any industry occurs through the entry of new firms and the growth of existing firms through increased market share. This market entry of new firms has its inherent costs. Some of these costs may arise due to various strategies of deterrence adopted by the existing firms. Other costs may be due to the market uncertainty, which may arise in the post-entry environment. The effects of entry include both the threat of potential entry into an oligopolistic industry regarding pricing, investment and R&D with advertising and also the actual entry affecting the strategies of oligopolistic suppliers in the post-entry game situation.

We analyze here the major costs and effects of market entry and their implications for the various strategies adopted by the incumbents and the new entrants. We discuss in some detail the role of sunk costs and market structure analyzed by the price competition model of Sutton (1991), which proposed that advertising and R&D spending can both be thought of as sunk costs along with investment costs for the acquisition of single plants of MES. Whereas the expenses of the acquisition of single plants of MES scale must be incurred by all new entrants and whose level is determined exogenously, the R&D and advertising outlays must be determined endogenously in industry equilibrium.

Secondly, we discuss the concept of "core competence" due to Prahalad and Hamel (1990) which is the collective learning in an organization comprising diverse production skills used to integrate multiple streams of technologies. It is the firm's cost competence that allows it to become more efficient than its competitors under a limited set of strategies and helps it to either enter a new industry or increase its market share in the existing industry.

In this framework we discuss the cost of the organizational slack termed by Leibenstein (1966) as "*X*-inefficiency." In particular we analyze the slack-ridden model of Selten (1986), which proposed a strong-slack hypothesis which maintains that slack has a tendency to increase so long as profits are positive; slack can be reduced but only under the threat of losses. The slack-based model of Selten has been used to compare the consequences of extremely weak ownership of firms with those of extremely strong ownership. This model predicts significant differences. In the usual theory of the symmetric, linear Cournot oligopoly with fixed costs, social welfare can be increased by restriction of free entry, if fixed costs are small but under the strong-slack hypothesis welfare can never be increased by restriction of entry. Free entry is always the best.

3.2 Sunk costs and entry

The two types of sunk costs distinguished by the Sutton model are exogenous and endogenous. The former involves setup costs or fixed outlays associated with the acquisition of a single plant of MES. The latter includes advertising, R&D outlays and any expenditure that raises consumers' willingness to pay for the firm's own product or services as opposed to rival products. The Sutton model incorporates sunk costs in industry equilibrium in terms of a two-stage game formulation. In the first stage of the game firms incur fixed plant expenditures for developing a product line. These expenditures are treated as sunk costs or exogenously given setup costs. In stage 2, price completion begins where the expectation of a tougher competitive regime makes entry less attractive, thus raising equilibrium concentration levels. If all firms produce a homogeneous product and the sunk costs are exogenous, then as the size of the market measured by the total number of firms entering the market increases due to increased profitability, concentration declines as the rates of market size to setup costs rises. But in the second stage the anticipation of tougher competition makes entry less attractive and thus raises the equilibrium concentration ratios. Sutton defines "the toughness of price competition" by a function linking the industry concentration to prices.

In case of endogenous sunk costs, for example advertising and R&D outlays, the game played at stage 1 may involve a competitive escalation of outlays by firms and hence higher sunk costs at the resulting equilibrium. Increase in market size may accentuate this process further.

The two-stage game in the Sutton model involving the decision about setup costs (i.e., the long-run costs) at stage 1 and the intensity of price competition at stage 2 (i.e., short-run decision) characterizes the post-entry equilibrium. For instance, if the degree of price competition at stage 2 is higher, the post-entry profits are lower and hence the entry is lower.

In this setup we discuss an example of the Sutton model in terms of Cournot competition, that is NE in quantities. We assume that each firm incurs a positive sunk cost on entering the industry and then produces at a constant marginal cost c.

In the symmetric Cournot case assume that n firms enter the market at stage 1. Then each firm i maximizes its own profit at stage 2

$$\pi_i = p_i\left(\sum_{j=1}^{n} x_j\right) x_i - cx_i \tag{3.1}$$

assuming its rivals' strategies as given. On setting $x_i = x$ for all i we obtain the Cournot equilibrium conditions

$$\begin{aligned} p &= c\left(1+\frac{1}{n\varepsilon - 1}\right) \\ x &= \frac{X}{n} = \frac{S}{pn} = \frac{S}{nc}\cdot\frac{n\varepsilon - 1}{n\varepsilon} \\ \pi &= (p-c)\,x = \frac{S}{(n^2\varepsilon)} \end{aligned} \tag{3.2}$$

where $\varepsilon = |((\mathrm{d}X/\mathrm{d}p)\,(\mathrm{p}/X))|$ is the absolute value of the price elasticity of demand, and we assume total output $X = \Sigma\ x_i$ as $X = S/p$ with S representing total consumer expenditure for this product. In case ε is set equal to one, these conditions become simpler, that is

$$\begin{aligned} p &= c\left(1+\frac{1}{n-1}\right) \\ x &= \frac{S}{c}\frac{n-1}{n^2} \\ \pi &= \frac{S}{n^2} \end{aligned} \tag{3.3}$$

Clearly for $n \geq 2$ we have $p > c$ and for $n \to \infty$ we have $p = c$ with zero profit. When a firm decides to enter the market with k rivals, its net profit is then $(S/(k+1)^2 - \sigma)$. So long as this net profit is positive, entry is profitable and hence in equilibrium one obtains $\pi = \sigma$ when there is no further entry. The equilibrium number of firms is then given by

$$n^* = \left(\frac{S}{\sigma}\right)^{1/2} \tag{3.4}$$

Thus if setup costs fall, the equilibrium number of firms increases. In the limit n^* becomes arbitrarily large if σ is zero (i.e., perfect competition).

Next, the Sutton model considers the role of endogenous sunk costs like advertising outlays, which may lead to product differentiation. As an example the model considers the Cournot model again with the perceived quality u_i for firm i, where each consumer is assumed to choose the good that maximizes u_i/p_i. If all firms offered the same quality level $\bar{u}$, then the equilibrium price level would be $p = (u/\bar{u})\,\bar{p}$, where $p = S/Q$, with Q denoting the total volume of output, and u is the quality level of the deviant firm. It follows

$$\bar{p} = S\left(\bar{Q} + \left(\frac{u}{\bar{u}}\right)q\right)^{-1} \tag{3.5}$$

where $\bar{Q}$ denotes the combined output of all nondeviant firms, while q is the output of the deviant firm. On using the price level $\bar{p}$ in the profit function of the nondeviant firms and then maximizing the profit function under the Cournot assumption, one obtains the common output level for the $(n-1)$ nondeviant firms as

$$\bar{q} = \left(\frac{u}{\bar{u}}\right)(n-1)\left[\left(\frac{u}{\bar{u}}\right)(n-1)+1\right]^2 \tag{3.6}$$

and a corresponding price

$$\bar{p} = c\left(1 + \frac{1}{\left(\frac{u}{\bar{u}}\right)(n-1)}\right) \tag{3.7}$$

The price, output and profit of the deviant firm in equilibrium would be

$$p(u|\bar{u}) = c\left(\frac{u}{\bar{u}} + \frac{1}{n-1}\right)$$

$$q(u|\bar{u}) = \left[(n-1) - \left(\frac{\bar{u}}{u}\right)(n-2)\right]\bar{q}$$

$$\pi(u|\bar{u}) = S\left(1 - \frac{1}{\frac{1}{n-1} + \frac{u}{\bar{u}}}\right)^2$$

Clearly, if all firms including the deviant set $u = \bar{u}$ in a symmetric fashion then we obtain $\pi = S/n^2$.

Now the Sutton model introduces the advertising outlay function $A(u)$ as

$$A(u) = \left(\frac{a}{\gamma}\right)(u^{\gamma} - 1), \quad \gamma > 1 \tag{3.8}$$

and the total fixed costs as

$$F(u) = \sigma + A(u) = \sigma + \left(\frac{a}{\gamma}\right)(u^{\gamma} - 1) \tag{3.9}$$

The elasticity of this total cost function is

$$E = \left(\frac{u}{F}\right)\left(\frac{\mathrm{d}F}{\mathrm{d}u}\right) = \gamma\left[1 - \frac{\left(\sigma - \frac{a}{\gamma}\right)}{F}\right]$$

which shows that as $\sigma < a/\gamma$ (or $\sigma > a/\gamma$) the elasticity $E >$ (or $<$)γ, where $E = \gamma$ when $u \to \infty$. We now specify a symmetric NE in advertising levels. The firm's profit level is

$$\pi(u|\bar{u}) - F(u)$$

where the firm sets a level u, where its rivals set a common value of perceived quality $\bar{u}$. On imposing the equilibrium marginal condition

$$\left(\frac{\mathrm{d}\Pi}{\mathrm{d}u}\right)_{u=\bar{u}} = \left(\frac{\mathrm{d}F}{\mathrm{d}u}\right)_{u=\bar{u}}$$

where all firms offer a common level of perceived quality $\bar{u} > 1$, one obtains

$$\frac{2S(n-1)^2}{n^3} = \gamma\left[F - \left(\sigma - \frac{a}{\gamma}\right)\right] \tag{3.10}$$

The value of F implicitly defined by Equation (3.10) is denoted by $F^*(n; S)$. Note that all those firms that enter the industry at stage 1 will in equilibrium set the same level of u and will all incur the same level of fixed outlays $F^*(n; S) \geq \sigma$. Hence the number of firms n will enter up to the point at which

$$\frac{S}{n^2} \geq F^*(n; S) \tag{3.11}$$

where n is the largest integer value. On combining the two Equations (3.10) and (3.11) as equality in equilibrium we obtain the equilibrium values of n and $F^*(n; S)$ as functions of σ, S, γ and a. The solution obtained by the Sutton Model takes the form

$$n + \left(\frac{1}{n}\right) - 2 = \left(\frac{\gamma}{2}\right)\left(1 - \frac{\sigma - \frac{a}{\gamma}}{F}\right) \tag{3.12}$$

Several comments on this model are in order. First of all, Equation (3.12) describes a locus in the (n, F) space and the equilibrium advertising level can be determined by the intersection of this locus with the zero-profit relation

$$F = \frac{S}{n^2}$$

This yields the result

$$\left(\frac{1}{n}\right) + (n-2) = \left(\frac{\gamma}{2}\right)\left[1 - \left(\sigma - \frac{a}{\gamma}\right)\frac{n^2}{S}\right]$$

indicating the dependence of the relative concentration C measured by $(1/n)$ and the market size S. Clearly, C falls monotonically as S rises if the condition $\sigma > a/\gamma$ holds, that is concentration falls as the market size S increases. In case $\sigma < a/\gamma$ the locus (Equation (3.12)) is downward-sloping and as S increases n first increases up to a point and then declines. For S tending to infinity, the equilibrium concentration C becomes independent of the setup cost but it still depends on γ, where $\gamma > 1$ is the parameter specified in the advertisement outlay function $A(u)$ defined by Equation (3.8).

Secondly, the role of advertisement expenditures A_i for firm i can be analyzed in the Cournot output model more directly by writing the profit function as

$$\pi_i = p(x)\, x_i - c_{X_i} - A_i \tag{3.13}$$

with total output X defined by total demand which depends on total advertisement outlays $A = \sum_{i=1}^{n} A_i$ in the industry. This is a symmetric case when all firms have identical marginal cost c and identical marginal response of advertisement expenditure $\partial x_i/\partial A_i = \partial X/\partial A$. The Cournot equilibrium in this simpler symmetric case is given by the optimal advertisement intensity $\lambda = A/pX$ as

$$\lambda = \frac{A}{pX} = \frac{\varepsilon_A}{n|\varepsilon_p|} \frac{1}{\beta(p(X)-c)-1} \tag{3.14}$$

where $\partial x_i/\partial A_i = \beta_i = \beta$. When $\beta = 1$ this takes the simpler form

$$\lambda = \frac{\varepsilon_A}{n|\varepsilon_p|} \frac{1}{p(X)-c-1} \tag{3.15}$$

Clearly the higher (or lower) price (advertisement) elasticity leads to lower (or higher) advertisement intensity. Also the price–cost relation can be written as

$$p = \left(c + \frac{1}{\beta}\right)\left(1 + \frac{1}{n|\varepsilon_p|-1}\right) \tag{3.16}$$

On using the relation $X = S/p$ as before where S represents total consumer expenditure for this industry, the concentration ratio $C = 1/n$ may be written as

$$C = \frac{\left[x\left(c + \frac{1}{\beta}\right)\left(1 + \frac{1}{n|\varepsilon_p|-1}\right)\right]}{S}$$

This shows that C falls as the market size S rises.

The optimal advertisement intensity condition (3.14) or (3.15), also known as the Dorfman–Steiner (1954) condition, has several interesting implications. First of all, if advertisement outlay does affect only demand X and not price, then we obtain a simpler condition from (3.15) as

$$\lambda = \frac{\varepsilon_A}{|\varepsilon_p|} \text{ or } \lambda_i = \frac{\varepsilon_{A_i}}{|\varepsilon_{p_i}|} \quad \text{(by symmetry)}$$

Thus in case of smaller price elasticity, each firm should spend more on advertising, that is low price elasticity leads to a higher advertising intensity. Secondly, one implication of the Dorfman–Steiner result is that if endogenous entry costs are important, the relation between market size and concentration should be flatter than when entry costs are exogenous. Cabral (2000) has presented empirical evidence for a series of 20 industries and 6 countries to support the hypothesis that the relation between concentration and size is flatter in case of endogenous entry costs (i.e., the plot is almost horizontal, whereas in case of exogenous entry costs the plot is downward sloping).

3.3 Sunk costs and limit pricing

Bain considered three sources of entry barriers: absolute cost advantage, product differentiation and scale economies. Absolute cost advantages refer to the situation that the firms already in the industry may have lower costs than those achievable by an entrant, for example patents, skilled personnel and technological expertise or higher capital cost for the entrants. Product differentiation refers to brand loyalty for the products of incumbent firms. Bain cited high levels of advertising to sales ratios as a major source of product differentiation barriers. With scale economies existing for the incumbent firms, the sales level needed to achieve cost parity with the incumbents may be sufficiently high so that the extra capacity introduced by the entrant will have the effect of depressing prices for everyone in the market. Thus while the previous price level would have been profitable for the entrant, the new price level is not. Note also that the scale economy of the incumbent may depend on not only the current output but cumulative output, that is the experience and accumulated skill levels. The role of capital expenditures as cumulated past investments is very critical in determining MES of incumbent firms. When capital expenditures once made become irreversible or "sunk" in the next period, an incumbent firm may be able to commit itself to producing an output that it could not sustain as an equilibrium if its first period capital expenditures were reversible. This precommitment to higher output discourages the established firm from cutting output in response to entry. This higher output by the incumbent acts as a deterrence to potential entry. This is the basic argument provided by the simple model of Dixit (1981) which showed that sunk costs allow an established firm to maintain a more aggressive response to actual entry than would be possible if costs were not sunk. This model allows production costs to depend on instilled capacity K in addition to output. Capacity which is measured in terms of output has a cost s per unit and the incumbent's cost function is assumed to be of the form

$$\begin{aligned} C(x,K) &= vx + sK + F \quad \text{for } x < K \\ &= (v+s)x + F \quad \text{for } x = K \end{aligned} \tag{3.17}$$

where x is output and F a reversible fixed cost. The potential entrant has no sunk cost like the incumbent and hence its cost function is of the form

$$C(x) = (v+s)x + F \tag{3.18}$$

Note that if there is excess capacity (i.e., $K > x$), the incumbent's marginal production cost is only v, whereas the entrant's marginal production cost is $v + s$, since he has built just enough capacity to produce its anticipated output. Thus the incumbent's marginal production cost advantage helps to deter entry in equilibrium. However, if capital expenditures were completely reversible, the established firm would have the same cost function and hence the same marginal production cost as the new entrant, and hence the incumbent firm would be unable to prevent entry. Thus Dixit's simple model supports Bain's structural view of economies of scale as a barrier to entry.

Dixit's model can be viewed as an extension of the limit pricing model originally developed by Bain (1956), Sylos-Labini (1962) and Modigliani (1958), which may be called the BSM model. The BSM model assumes that the established firms would adopt an aggressive strategy perhaps by cutting prices before entry occurs, but Dixit's model extends this view of limit pricing by building the credibility aspect of entry deterrence by the incumbent firm. In this model of credible limit pricing the limit output (i.e., output corresponding to the limit price) must be an equilibrium output of the post-entry game. But if demand is not sufficiently elastic, the established firm cannot maintain an equilibrium output large enough to deter entry.

The empirical evidence for the hypothesis that excess capacity by the incumbents acts as a deterrence to new entry is rather mixed. The cross-section estimates by Masson and Shaanan (1983) show rather weak evidence in this regard. On the role of endogenous sunk costs Kessides (1986) estimated the degree to which advertising expenditures are sunk in a sample of 266 four digit US industries between 1972 and 1977 and found that advertising-based sunk costs significantly reduces entry rates. Moreover advertising-based sunk costs seemed to be more important than those associated with physical capital. Geroski *et al.* (1990) have discussed in some detail an empirical study by Gilbert (1986) who used data on MES and factor costs in 16 industries to investigate the conditions under which entry prevention may be credible by a single incumbent firm. The larger is the MES of entry and the larger is the share of costs that are sunk by an established firm, the easier the entry prevention. Dixit's model predicts that the maximum output sustainable against entry increases with the share of total sunk costs and the MES of entry and the elasticity of demand. Given the share of sunk cost and the MES of entry the required elasticity of demand for a single established firm to prevent entry can be calculated. For the industries studied by Gilbert (1986) the required demand elasticity must be at least

1.5 (e.g., for cement) and the demand elasticity varies from 2.0 up to 4.3 (for petroleum refining) for the 16 industries. But this is long-run elasticity. A short-run market elasticity may be much lower than 2.0, in which case a single firm in Gilbert's sample would have a difficult time maintaining a large enough output to exclude entry.

Advertising may be combined with the scale-related barrier to entry, so that the established firms might find it profitable to enhance product differentiation in order to exploit the benefits of scale economies. Dixit (1979) developed a simple model to combine the three types of entry barriers, for example product differentiation, scale economies and other means of entry deterrence. The two firms, firm 1 being the incumbent and firm 2 the potential entrant, have the linear demand function

$$p_i = a_i - b_i x_i - c x_j, \quad i \neq j; \quad i, j = 1, 2 \tag{3.19}$$

with a_1, b_1 being positive and $c < (b_1 b_2)^2$. Each firm has a cost function $C_i = mx_i + F$. It is assumed that firm 1 acts as a Stackelberg leader. The limit output in this model is

$$Y = \frac{[a_2 - m - 2(b_2 F)^{1/2}]}{c} \tag{3.20}$$

It is clear that the extent of product differentiation affects all the parameters of demand for the two brands. For instance, an increase in advertising expenditures that enhances the demand for brand 1 at the expense of brand 2 may increase a_1 relative to a_2 and decrease the marginal cost c. The decrease in the cross-product term c lowers the cross-elasticity of demand for the two brands, that is they become poorer substitutes. The effects of all demand-increasing activities can be examined by comparing profits earned when entry occurs to the profits earned at the limit output given by Equation (3.20). For example, a decrease in a_2 lowers the limit output, while a decrease in c has the opposite effect.

Note that production-differentiation barriers in the form of high levels of advertising intensity occur simultaneously with the scale barriers due to scale economies in production, advertising and R&D spendings and reaction barriers posed by high levels of industry concentration because of the threat of oligopolistic coordination in the face of entry. Seller concentration has two opposite effects on the likelihood of entry: reaction barrier and inducement. Thus if an entrant can survive the reaction barrier posed by high seller concentration, it too can enjoy the less rivalrous condition of such an industry.

While concentration is one indicator of the severity of rivalry, another is the importance of price as the basis of competition, also called the toughness of price competition in the Sutton model. Industries in which price is not the major basis of competition should be more attractive for new entrants, since they usually offer greater stability in profits and less risk of a reaction barrier for the new entrant.

3.4 Cost of slack in oligopoly

We consider in this section the role of sunk costs in the form of slack-ridden costs under oligopoly. We discuss in some detail the model of Selten (1986) and its general implications for entry dynamics.

For each firm i ($i = 1, 2, \ldots, n$) we assume a linear demand and cost function as

$$p = b - aX \quad \text{and} \quad C(x_i) = F + cx_i + s_i x_i$$

where x_i is output and X the total industry output with F as fixed cost and c the constant marginal cost assumed to be equal for all firms. Profits are then

$$\pi_i = px_i - cx_i - F - s_i x_i \tag{3.21}$$

where $s_i x_i$ is the slack-ridden cost which has a tendency to rise whenever profits rise.

On maximizing π_i for each i with respect to s_i, $0 \leq s_i \leq 1$, for fixed $x_i > 0$ we obtain the optimal slack-ridden cost s_i which equates marginal revenue to the marginal cost from slacks. Assuming an interior solution for s_i we maximize π_i with respect to x_i and obtain the symmetric Cournot equilibrium as follows:

$$\begin{aligned} X &= \sum_{i=1}^{n} x_i = \frac{b-c}{a}\,\frac{n}{1+n} - \frac{n\bar{s}}{a(n+1)}; \quad \bar{s} = \left(\frac{1}{n}\right)\Sigma s_i \\ x_i &= \left(\frac{1}{a}\right)(b - c - s_i) - X \\ \pi_i &= \left(\frac{1}{a}\right)(A_1 - A_2)^2 - F \end{aligned} \tag{3.22}$$

where

$$A_1 = b - c - s_i, \quad A_2 = (b - c - s_i) - \frac{n}{1+n}(b - c - \bar{s})$$

Selten uses normalized values as $a = b - c = 1$ as an example. This yields

$$\pi_i = \left(\frac{1}{1+n} + \frac{n\bar{s}}{1+n} - s_i\right)^2 - F$$

In long-run equilibrium, $\pi_i = 0$ for all i, hence

$$\left(\frac{1}{1+n} + \frac{n\bar{s}}{1+n} - s_i\right)^2 = F \tag{3.23}$$

The left-hand side is a decreasing function of s_i and hence for given $\bar{s}$ there is at most one s_i which satisfies Equation (3.23). This shows that in long-run equilibrium all slack rates are equal, that is $s_i = s$ for all i. This yields

$$\frac{(1-s)^2}{(n+1)^2} = F \tag{3.24}$$

or

$$s = 1 - (1+n)\sqrt{F}$$

Since $0 \leq s \leq 1$ we have from Equation (3.24)

$$\frac{1}{(n+1)} \geq \sqrt{F} \tag{3.25}$$

Let n^* be the maximum number of competitors and $[n^*]$ be its integral value. Then

$$n^* = \max_{n=1,2,\ldots} \left\{ n \left| \frac{1}{n+1} \right| \geq \sqrt{F} \right\}$$

and $[n^*]$ is the number of firms in industry equilibrium. Note that entry restrictions through government regulations may result in the number of competitors smaller than $[n^*]$. Thus if there are $n-1$ firms in the industry, and one potential competitor denoted by the n-th firm with a zero slack considers entry, then the entry will be successful if and only if $n \leq n^*$. For $n > n^*$ he has to expect long-run losses.

Selten also derives the social welfare consequences in this industry. Denoting by W_n the total of consumers' surplus (CS) and producers' surplus (PS) and a proportion of total slack S, that is $W_n = \text{CS} + \text{PS} + \alpha\text{S}$, where n is the number of firms and $D_n = W_n - W_{n-1}$, he derives the following conditions:

$$W_n = \left(\frac{n^2F}{2}\right) + \alpha n\sqrt{F} - \alpha n(n+1)F$$

$$D_n \geq F\left[(1-\alpha)\, n - \left(\frac{1}{2}\right) + \alpha\right]$$

In view of $0 \leq \alpha \leq 1$ and $n \geq 1$, this yields

$$D_n \geq \left(\frac{1}{2}\right) F \tag{3.26}$$

This shows that social welfare W_n is always increased by the entry of a new entrant so long as the maximal number $[n^*]$ of firms has not yet been reached. Thus restriction of entry cannot increase social welfare when there is positive slack. In case of zero slack, however, one could derive

$$D_n = [2n(n+1)]^{-1}\left\{\frac{1}{n+1} + \frac{1}{n}\right\} - F$$

which shows that D_n is a decreasing function of n. Hence social welfare can be increased by restrictions on entry for $n < n^*$.

Two general implications of the Selten model may be discussed here. First of all, in case of the general slack hypothesis it can be shown that

$$\text{CS} = \frac{n^2F}{2}$$

implying that higher F may yield higher CS. Selten introduces the dynamics of slack as an adjustment process:

$$\dot{s}_i = \frac{ds_i}{dt} = \mu_1 r_i - \mu_2 s_i \quad \text{for } s_i > 0$$

where r_i is the profit rate and μ_1, μ_2 are nonnegative constants. This adjustment process assumes that the slack rate s_i will increase more quickly if the profit rate r_i is high; it will decrease when the level of slack rate is already high, since with higher slacks the inefficiency is more visible. This shows that the slack rates have important effects on the efficiency of firms in the industry.

Secondly, the slack-ridden costs may also be introduced in a general Pareto efficiency model, where a firm k chooses the optimal output x by maximizing profits

$$\pi = \{p - (c + s_k)\}\, x - F_k$$

as follows:

$$\text{Max } \pi$$

$$\text{subject to } \sum_{j=1}^{n} z_{ij}\lambda_j \leq z_{ik}, \quad i = 1, 2, \ldots, n \tag{3.27}$$

$$\sum_{j=1}^{n} x_j\lambda_j \geq x; \quad \Sigma\lambda_j = 1,\ \lambda_j \geq 0,\ j = 1, 2, \ldots, n$$

Here there are m inputs (z_{ij}) and one output (x_j) for each firm j and the cost function is $C = (c + s_k)\, x + F$. Here the economies due to R&D may be analyzed. If $s_k = 0$ for all k, and $p = a - bX,\ X = \Sigma x_j$ then the optimal conditions for an efficient output are

$$\begin{aligned} &(a - c) - bX^* - bx^* - \alpha^* \leq 0 \\ &\alpha^* x_k - \beta^{*\prime} z_k - \alpha_0^* \leq 0 \\ &\alpha^*,\ \beta^* \geq 0, \quad \alpha_0^* \text{ free in sign,} \end{aligned} \tag{3.28}$$

where z_k is the input vector of firm k. When firm k is efficient, then equality must hold in the first two restrictions of Equation (3.28). This yields

$$x^* = \frac{a - c}{b} - X^* - \frac{\alpha^*}{b}$$

If $n_1 < n$ firms are efficient, then

$$X_{n_1}^* = \left(\frac{n_1}{b}\right)(1 + n_1)^{-1}\ (a - c - \alpha^*)$$

where $X_{n_1} = \sum_{j=1}^{n_1} x_j$. In case of the slack-based costs one could easily derive

$$X_{n_1}^* = \left(\frac{n_1}{b}\right)(1 + n_1)^{-1}\ (a - c - \bar{s} - \alpha^*)$$

where $\bar{s}$ is the average rate of slack. Profits are then

$$\pi_k^* = (p - c - s_k)\, x^* + F$$

$$p = a + n_1(1 + n_1)^{-1}\,(\alpha^* + \bar{s} - a - c)$$

By imposing the zero-profit condition in the long run, $\pi_k^* = 0$, for all efficient firms, one would obtain the equilibrium price as

$$p = c + \bar{s} + \left(\frac{F}{x^*}\right)$$

This shows that when the new entrant firms are efficient, it tends to decrease price. Thus efficiency like advertisement may have a demand enhancing effect. This model (Equation (3.27)) also shows that slacks may exist on both input and output sides. The slack variables in the LP model (Equation (3.27)) may be related to the concept of slack introduced by the Selten model.

3.5 The effects of entry

We discuss in this section two models analyzing the effects of entry on industry performance. One is the model of Seade (1980) analyzing the market growth in an oligopoly framework and the other of Spence (1984) analyzing the effect of cost reduction through R&D and spillover of knowledge in a competitive industry framework.

Seade (1980) discusses the effects of entry in three aspects: (a) effects on outputs and profits; (b) comparative statics of the equilibrium and (c) the firms may have a certain degree of collusion, which is called a "quasi-Cournot" world. The model assumes a homogenous output Y produced by n firms with outputs and the inverse demand function $p = p(Y)$. The profit for firm i is

$$\pi_i = p(Y)\, y_i - c_i(y_i)$$

and the profit-maximization condition is

$$\frac{\mathrm{d}\Pi_i}{\mathrm{d}\partial y_i} = p + y_i p' \left(\frac{\mathrm{d}Y}{\mathrm{d}y_i}\right) - c_i' = 0$$

where $\mathrm{d}Y/\mathrm{d}y_i$ is the conjectural variation. The model assumes this to be a positive constant λ_i representing a reflection of each firm to protect its market share. Assuming identical firms and a symmetric equilibrium, one obtains

$$p+\lambda y p' = 0 \tag{3.29}$$

and

$$\lambda^2 y p'' + 2\lambda p' - c'' < 0$$

where prime denotes differentiation with respect to y_i and all subscripts are dropped by the symmetry assumption. Seade (1980) allows the number of firms n to be a continuous variable on which everything depends differentially.

On differentiating the first equation of Equation (3.29) with respect to n, one obtains

$$p'\left(n\frac{\mathrm{d}y}{\mathrm{d}n}+y\right)+\lambda y p''\left(n\frac{\mathrm{d}y}{\mathrm{d}n}+y\right)+\lambda p'\frac{\mathrm{d}y}{\mathrm{d}n}=c''\frac{\mathrm{d}y}{\mathrm{d}n} \tag{3.30}$$

Since $p' < 0$, Equation (3.30) yields

$$\varepsilon_{yn} = -\frac{(E+m)}{(E+m+k)} \tag{3.31}$$

where ε_{yn} is the elasticity of demand y with respect to n, and $m=n/\lambda$, $E=Yp''/p'$ and $k=1-c''/(\lambda p')$. The model interprets m as the number of "effective" firms in the industry and it is assumed that $1 \leq m \leq n$ and k is positive since $k<0$ implies $c'' < \lambda p'$ (i.e., falling marginal costs). The second equation of Equation (3.29) implies then

$$E+m+mk > 0 \tag{3.32}$$

The equilibrium then is stable for $E > -(m+k)$ and a saddle point for $E < (m+k)$. When the stability condition holds in the form, that is $E > -m$ the model gets the result $\varepsilon_{yn} < 0$, which was derived by Ruffin (1971) before. The proof follows from the fact that given $E+m > 0$ the condition $E+m+k \leq 0$ implies $k<0$, which in turn implies $E+m+mk \leq 0$ given $m \geq 1$, which contradicts Equation (3.32). Hence it must hold that $E+m+k > 0$ and by Equation (3.31) it yields that the elasticity ε_{yn} is negative, that is $\mathrm{d}y/\mathrm{d}n < 0$.

From the definition of total output $Y=ny$ one obtains

$$\varepsilon_{Yn} = \frac{k}{(E+m+k)} \tag{3.33}$$

Hence if k and $(E+m)$ are both positive, this yields $\varepsilon_{Yn} > 0$, that is $\mathrm{d}Y/\mathrm{d}n > 0$. Thus total output always expands as entry into stable equilibria occurs. Likewise the condition

$$\frac{\mathrm{d}\Pi_\mathrm{T}}{\mathrm{d}n} = \frac{\mathrm{d}(n\Pi_i)}{\mathrm{d}n} = n\frac{\mathrm{d}\Pi_i}{\mathrm{d}n} + \Pi_i$$

where Π_T = total industry profits, yields $\mathrm{d}\Pi_\mathrm{T}/\mathrm{d}n < 0$, when profit for each firm declines as n rises since

$$\frac{\mathrm{d}\Pi_i}{\mathrm{d}n} = \frac{y^2 p'}{m}\left[\frac{E+m+mk}{E+m+k}\right]$$

this is because $p' < 0$ and due to the assumption in Equation (3.32) and as $E > -(m+k)$.

The effect of λ which represents a variable degree of collusion one could also derive the results

$$\varepsilon_{y\lambda} = -\frac{1}{(E+m+k)} \tag{3.34}$$

and

$$\frac{\mathrm{d}\Pi_i}{\mathrm{d}\lambda} = -y^2 p' \frac{(m-1)}{(E+m+k)}$$

$$\varepsilon_{Y\lambda} = \varepsilon_{y\lambda} \text{ and } \frac{\mathrm{d}\Pi_\mathrm{T}}{\mathrm{d}_\lambda} = n\,\frac{\mathrm{d}\Pi_i}{\mathrm{d}\lambda}$$

Under the assumption that $(E+m+k)$ is positive (i.e., the case of stable equilibrium) it follows that both y and Y would fall, while Π_i and Π_T would rise with λ. Thus collision always works so as to increase individual profits.

The effects of entry under conditions of Arrow's learning by doing are analyzed by Jovanovic and Lach (1989). Following Arrow, it is assumed that learning is a function of cumulative gross investment and that technical changes resulting from this learning are completely embodied in new capital goods. Let z_t be cumulative gross investment as the number of machines installed before time t. Each machine has a fixed installation cost $k(z_t)$ and a variable unit cost of production $c(z_t)$, where it is assumed that $c' = \partial c/\partial z < 0$ and $k' < 0$. Equilibrium is defined by the condition that firms take the time path of prices as given and each entrant earns zero discounted profits, that is

$$\int_\tau^\infty \mathrm{e}^{-r(s-\tau)} \max\left\{0, [P(z_t) - c(z_t)]\right\} \mathrm{d}s - k(z_\tau) = 0 \tag{3.35}$$

Here $P(z_t)$ is the market clearing price, where $p_t = P(z_t)$, p_t being the solution to the demand $D(p_t)$ equaling supply $z_t - x_t$, x_t being the cumulative exit up to time t:

$$z - x(p) = D(p) \tag{3.36}$$

where

$$x_t = c^{-1}\{\min[p_t,\ c(0)]\} = x(p_t) \tag{3.37}$$

It follows the Equation (3.37) for x_t that $\dot{x} = \mathrm{d}x/\mathrm{d}t = x'P'\dot{z}$. On differentiating both sides of Equation (3.36) one obtains

$$P' = \frac{1}{(D' + x')}$$

and hence

$$\dot{x} = \frac{x'}{D' + x'}\dot{z} < \dot{z}$$

Clearly, $x' = 0$ when the market clearing price P is above $c(0)$, that is when z is small so that $\dot{x} = 0$. Note also that net entry defined by $(\dot{n} - \dot{x})$ is always positive since

$$\dot{n} - \dot{x} = \frac{D'}{D' + x'}\dot{z} > 0 \tag{3.38}$$

Thus the number of firms and the total industry output, $z - x = z - x[P(z)]$, grows as long as cumulative gross investment increases (i.e., $\dot{z} > 0$).

Jovanovic and Lach show that new entry may be viewed as the use of new technology, for example a new type of software through R&D investment, then z_t may represent the cumulative diffusion of new technology. An example of this diffusion path is given in the model as follows. Assume the case of constant elasticity of demand

$$k(z) = Az^{-\alpha}, \quad D^{-1}(z) = Bz^{-\beta}; \quad 0 < \alpha < \beta < 1$$

For equilibrium the total gross entry N must be such that any further entry would yield negative discounted profits defined by Equation (3.35). Hence

$$r^{-1}[P(N) - c(N)] - k(N) = 0 \tag{3.39}$$

and the following equation can be derived from Equation (3.35).

$$\dot{z} = \frac{P(z) - c(z) - rk(z)}{-\{k'(z) + r^{-1}c'(z)\}\{1 - e^{-rL_\tau}\}} \tag{3.40}$$

when $L_\tau = T_\tau - \tau$ is the lifetime of vintage τ. The solution of Equation (3.40) subject to the initial condition $z_0 = 0$ is

$$z_t = N(1 - e^{-\lambda rt})^{1/(\lambda\alpha)} \tag{3.41}$$

Solving Equation (3.39) for N yields

$$N = \left(\frac{B}{rA}\right)^{1/\lambda\alpha}; \quad \lambda = \frac{(\beta - \alpha)}{\alpha} > 0 \tag{3.42}$$

Thus the diffusion path $\{z_t;\ 0 \le t \le \infty\}$ can be easily derived from Equations (3.41) and (3.42) as

$$\dot{z} = \frac{Bz_t^{1-\lambda\alpha}}{\alpha A} - \frac{rz_t}{\alpha} > 0$$

This is close to an S-shaped curve in equilibrium. This learning-by-doing model also exhibits a strictly increasing industry output with the diffusion through learning by doing and a strictly decreasing product price and the rate of profit for each vintage of entrants.

The learning curve effects are utilized by Cabral and Riordan (1994) in a different setting where priced-setting-differentiated duopolists are selling to a sequence of heterogeneous buyers with uncertain demand. In each period a buyer demands at most one unit of the good from one of the two firms. It is assumed that a sale always occurs and x is the premium a buyer is willing to pay for firm 2's product. The preference parameter x is assumed to be identically and independently distributed across buyers with a cumulative distribution function $F(x)$, whose density $f(x)$ is assumed to be differentiable and symmetric about zero such that $H(x) = F(x)/f(x)$ is increasing. Let $C = c_2 - c_1$ be the cost difference of two firms. Subtracting firm 1's first-order condition from the corresponding condition for firm 2 yields $P + G(P) = C$, where $G(x) = H(x) - H(-x)$. This is an equilibrium condition representing the price difference as a function of cost difference. Equilibrium-expected profit or "one show profit" is then $\Pi(P) = \mathrm{H}(P) \cdot \mathrm{F}(p)$, where $P = p_2 - p_1$ and $F(P)$ is the probability that firm 1 makes a sale. Note that the first-order condition for firm 1 is $p_1 - H(p) = c_1$.

To represent the learning curve effect Cabral and Riordan assume that a firm's unit cost $c(s)$ is a decreasing function of cumulative past sales (s) and that after m sales there is no further learning (i.e., the firm reaches the bottom of the learning curve). Under these assumptions the firms are maximizing expected discounted profits. The solution concept used by the authors is Markov perfect equilibrium (MPE) with the property that each firm's strategy depends only on the state of the game. The state of the game is defined by a pair (i, j), where i and j are the cumulative sales of firms 1 and 2 respectively. Given a strategy for each firm they define a value function $v(i, j)$ giving the value of the game for firm 1 in state (i, j). Then they derive two important results:

$$p(i,j) - H(-P(i,j)) = c(i) - \delta w(i,j) \tag{3.43}$$

and

$$v(i,j) = \pi(-P(i,j)) + \delta v(i, j+1) \tag{3.44}$$

where δ is the discount factor and $w(i,j) = v(i+1, j) - v(i, j+1)$, that is the "prize" from winning a sale and firm 1's strategy is given by

$$p(i,j) - P(i,j) \quad \text{with } P = p_2 - p_1$$

The first result is the first-order condition for the symmetric one-shot game except that the discounted "prize" from winning enters as a production subsidy. Hence the difference in prices equals the one-shot equilibrium price difference corresponding to these subsidized costs, that is

$$P(i,j) + G(P(i,j)) = C(i,j) - \delta W(i,j)$$

where $C(i,j) = c(i) - c(j)$ and $W(i,j)$ is the difference between firm 1's and firm 2's prize.

Now we discuss a model due to Spence (1984) which explores the effects of cost-reducing investments (i.e., R&D expenditures) by firms. The model assumes n firms indexed by i with $c_i(t)$ as unit costs, where

$$c_t(t) = F(z_i(t))$$

where $z_i(t)$ is the accumulated knowledge with respect to cost reduction. Let $m_i(t)$ be the current spending by firm i on R&D. Then it is assumed that

$$\dot{z}_i(t) = m_i(t) + \theta \sum_{j \neq i} m_j(t) \tag{3.45}$$

where $m_i(t)$ is the current expenditures by firm i on R&D and the parameter θ captures spillover (i.e., $\theta = 0$ denotes no spillovers). The accumulated investment is then

$$z_i = M_i + \theta \sum_{j \neq i} M_j$$

where

$$M_i = \int_0^t m_i(s)\, ds$$

The benefits in dollars from the sale of x units of the good are $B(x)$. The inverse demand is $B'(x)$ and the profit of firm i is

$$E_i = x_i B'(x) - c_i x_i$$

where $x = \Sigma x_i$, $x_i =$ output of firm i. It is assumed that there is a unique equilibrium at each time point in the market, for example an NE. Let $x_i(z)$ and $x(z) = \Sigma_i x_i(z)$ be the equilibrium. The CS is then $B(x(z)) - xB'(x(z)) = H(z)$. The earnings gross of R&D expenditures for firm i are

$$E_i(z) = x_i(z)\, B'(x(z)) - c_i(z_i)\, x_i(z)$$

The present value of its earnings net of R&D investment is

$$V_i = E_i(z) - (1 - s)\, M_i \tag{3.46}$$

where there is a subsidy of s for R&D, so that each dollar of R&D costs the firm $(1 - s)$. The firm i takes the M_j of its rivals as given and then maximizes V_i with respect to M_i by setting

$$E_i^i + \theta \sum_{j \neq i} E_j^i = (1 - s) \tag{3.47}$$

The solution to these n equations is the market equilibrium, where E_j^i is the derivative of E_i with respect to z_j. Given the equilibrium values of $M = (M_1, M_2, \ldots, M_n)$ the performance of the market can be evaluated in terms of the total surplus

$$T(M) = H(z(M)) + \sum_i V_i(M) - x \sum_i M_i$$

where the last term reflects the costs to the public sector of the R&D subsidies, and $H(z(M))$ is the CS. They consider a simple example in a symmetric case of NE where demand is of the constant elasticity variety, that is $x = Ap^{-b}$ with unit costs $c = q_0 + c_0\, e^{-gz}$. It is assumed that a unique static NE in quantities exists. Let $w = 1 - 1/nb$, where n is the number of firms and b is the price elasticity of demand. The earnings for a firm is then

$$E(z, n) = Ab\left[\frac{w^{b-1}}{n}\right](1 - w)c^{1-b}$$

The CS is

$$H(z, n) = \left[\frac{A}{b-1}\right]w^{b-1}c^{1-b}$$

and the total surplus is

$$T(z, n, \theta) = Aw^{b-1}\left[\frac{1}{b-1} + 1 - w\right]c^{1-b} - \left(\frac{n}{K}\right)z$$

where $K = 1 + \theta(n-1)$. The market minimizes the function

$$Q = R(z) - (1 - s)\, z$$

where

$$R(z) = \frac{A}{n(b-1)}\, w^{b-1} \cdot \left[2w + \frac{K^{(b-1)(1-w)}}{n} - 2w\right]c^{1-b}$$

Note that the coefficient of K/n in $R(z)$ is negative. Thus an increase in θ increases K for $n > 1$ and hence reduces R and $R_z = \partial R/\partial z$. Thus other things being equal any increase in knowledge spillovers reduces cost reduction.

Note that other costs at the industry level are given as

$$\text{R\&D} = \frac{zn}{K} = zn[1 + \theta(n-1)]^{-1}$$

If θ is zero the costs are proportional to the number of firms. With θ positive the unit costs have an upper limit of $1/\theta$ as n increases. Thus while spillovers reduce the incentives for cost reduction, they also reduce the costs at the industry level of achieving a given level of cost reduction. Thus although spillovers have a significant negative effect on cost reduction, it may be a mistake to completely eliminate it. Spence has derived the important result that the markets characterized by high spillover but populated by firms which underestimate the spillovers on their effects on prices perform much better than the same markets populated by fully informed firms. More aggressive R&D investments based on underestimated spillovers increase entry barriers and reduce the number of active competitors. His conclusion is as follows: "The fact that the producer of knowledge does not appropriate all the social benefits leads to the conclusion that it should at least be rewarded for the benefits it confers on other firms. And even that falls short of the social benefits. But it does not follow that other firms should pay for it. If they do pay for it, they are paying more than its marginal cost. There is a direct analogue with public goods. The output of R&D has the character of a public good. The incentives are weak for individuals to supply it. But we do not approach the solution to public goods problems by contriving to have the beneficiaries pay for it where possible, because that leads to under consumption and suboptimal use."

3.6 Concluding remarks

The various entry models discussed above basically assume that profit signals induce entry flows. But the rational entrants will not be attracted by pre-entry profitability so much as post-entry profitability. Kessides (1986) has emphasized the post-entry situation in his models much more than the pre-entry framework. His starting point is to assume that the number of new firms N_e that are observed is such that the expected post-entry profits are zero. Thus assume a homogeneous good with a pre-entry price p_0, where the efficient scale of entry (output) is x. Then the arrival of N_e entrants results in a post-entry price p_e of approximately

$$p_e = p_0 \frac{(1 - N_s s)}{\varepsilon} \tag{3.48}$$

where ε is the absolute value of price elasticity of demand and $s = x/q_0 =$ the ratio of the efficient scale of entry to pre-entry output. Hence the change in price due to entry is

$$\Delta p = p_e - p_0 = -p_0 \frac{(N_e s)}{\varepsilon} \tag{3.49}$$

Furthermore, if the market grows at a positive rate g and the incumbents at a rate r, then Equation (3.49) can be written as

$$\Delta p = -p_0 \frac{[N_e s - (g - r)]}{\varepsilon(1+g)} \tag{3.50}$$

The relation shown in Equation (3.50) exhibits a number of useful implications. First of all, if the market grows fast enough, then the post-entry price could actually exceed pre-entry prices. Secondly, a more elastic demand yields a smaller price change, as does a smaller efficient scale of entry and also a smaller number of entrants. Note that advertising enters in this model as a sunk cost of entry. Kessides applied this model to 266 four-digit US industries in 1972–1977 and found that advertising leads to a sunk cost barrier to entry, which is more important than physical capital deterring entry. He has also found that the entrants perceive a greater likelihood of success in highly advertising-intensive industries.

A second comment on the various models discussed so far deals with the point that they do not apply to the new markets. According to Geroski (2003) why so much entry occurs so early on in the life of a new market is puzzling. The profitability leading to entry does not apply here as much, since very young markets have very few consumers. One reason is the lack of substantive barriers to entry, that is the setup costs are very small and the new entrants also are small. Early consumers are also very price sensitive. A second reason may be the effect of new technology and new markets on the bundle of possibilities. The technological opportunities in new markets or new technologies may lead to other rich possibilities later on. Thus it may be much easier for an entrant to develop new and related products. Also the future growth possibility of new markets provides an important attraction.

References

Bain, J. (1956) *Barriers to New Competition* (Cambridge: Harvard University Press).

Cabral, O.M. (2000) *Introduction to Industrial Organization* (Cambridge: MIT Press).

Cabral, O. and M. Riordan (1994) The Learning curve, market dominance and predatory pricing, *Econometrica* 62, 1115–1140.

Dixit, A. (1979) A model of duopoly suggesting a theory of entry barriers, *Bell Journal of Economics* 10, 20–32.

Dixit, A. (1981) The role of investment in entry deterrence, *Economic Journal* 90, 95–106.

Dorfman, R. and P. Steiner (1954) Optimal advertising and optimal quality, *American Economic Review* 44, 826–836.

Geroski, P.A. (2003) *The Evolution of New Markets* (Oxford: Oxford University Press).

Geroski, P., R. Gilbert and A. Jacquemin (1990) *Barriers to Entry and Strategic Competition* (London: Harwood Academic Publishers).

Gilbert, R. (1986) Preemptive competition, in J. Stiglitz and G. Mathewson (eds) *New Developments in the Analysis of Market Structure* (Cambridge: MIT Press).

Jovanovic, B. and S. Lach (1989) Entry exit and diffusion with learning by doing, *American Economic Review* 79, 690–699.
Kessides, I. (1986) Advertising, sunk costs and barriers to entry, *Review of Economics and Statistics* 68, 84–95.
Leibenstein, H. (1966) Allocative efficiency versus X-efficiency, *American Economic Review* 66, 392–415.
Masson, R. and J. Shaanan (1983) Social costs of oligopoly and the value of competition, *Economic Journal* 92, 80–89.
Modigliani, F. (1958) New developments in oligopoly front, *Journal of Political Economy* 66, 215–232.
Prahalad, C.K. and G. Hamel (1990) The core competence of the corporation, *Harvard Business Review* 66, 79–91.
Ruffin, R. (1971) Cournot oligopoly and competitive behavior, *Review of Economic Studies* 38, 493–502.
Seade, J. (1980) On the effects of entry, *Econometrica* 48, 479–490.
Selten, R. (1986) Elementary theory of slack-ridden imperfect competition, in J. Stiglitz and G. Mathewson (eds) *New Developments in the Analysis of Market Structure* (Cambridge: MIT Press).
Spence, M. (1984) Cost reduction, competition and industry performance, *Econometrica* 52, 101–122.
Sutton, J. (1991) *Sunk Costs and Market Structure* (Cambridge: MIT Press).
Sylos-Labini, P. (1962) *Oligopoly and Technical Progress* (Cambridge: MIT Press).

4
Entry and Market Structure

4.1 Introduction

The dynamics of entry behavior and its impact depend very critically on the type of market structure. If an industry has a large number of competitors and the market is competitive the Walrasian model of entry behavior provides an adequate approximation of the market evolution. However, if the number of firms is small and the market is oligopolistic, the entry behavior follows the strategies of a two-stage or multistage game between the incumbents and the potential entrants. There exist other types of market structures where the dynamics of entry behavior depend on the diffusion of a new technical innovation or a new product. Also in international markets entry occurs through joint ventures.

Our objective here is threefold. First, we analyze the economic implications of industry concentration for the dynamic entry behavior. Secondly, we consider the innovation processes in the Schumpeterian framework as a generalized view of the entry process. This is analyzed more specifically through the R&D process affecting the entry behavior. Finally, we analyze entry in foreign markets where the dynamics of comparative advantage and comparative efficiency play significant roles.

4.2 Market size and concentration

A simple way to relate entry to market size is to model firm behavior in equilibrium. Let $p = a - X/S$ be the inverse demand function and $C_i = cx_i + F$ is the cost function for firm i with $X = \sum_{i=1}^{n} x_i$ as total industry output. Firms are assumed to be of the same size and the good is a homogenous product. Here S is a measure of the market size. Assuming a symmetric equilibrium and a classical Cournot framework it is easy to derive the equilibrium profit for each firm as

$$\pi(n) = S\left(\frac{a-c}{1+n}\right)^2 - F \tag{4.1}$$

with total industrial output and price as

$$X(n) = \frac{(a-c)S}{1+n}, \quad p(n) = \frac{a-n(a-c)}{(1+n)} \tag{4.2}$$

Note that total demand $D=X$ may be written as $D=(a-p)S$, where the market size S is usually represented by total consumer expenditure for the product. Clearly as S rises, total demand rises and the Equation (4.1) may be used to define a free-entry equilibrium, that is the equilibrium number of firms n^* is given by the zero-profit condition

$$n^* = (a-c)\sqrt{\frac{S}{F}} - 1 \tag{4.3}$$

such that no potential entry occurs (i.e., $\pi(n^*+1) \leq 0$) and no firm wishes to exit (i.e., $\pi(n^*-1) \geq 0$). Hence the equilibrium value of n is given by $[n^*]$ which denotes the highest integer lower than or equal to n^*. Note that the number of active firms $[n^*]$ in equilibrium varies less than proportionately with respect to market size S. This is due to increased price competition as n^* rises and the market tends to the competitive direction.

Industry concentration related to market size is measured in several ways. The most common measure is the coefficient C_m which is the sum of the market shares (s_i) of the largest m firms, that is

$$C_4 = \sum_{i=1}^{4} s_{(i)}, s_{(1)} > s_{(2)} > s_{(3)} > \ldots > s_{(m)}$$

where firms are ordered by market shares. A second measure of market concentration is the Herfindahl index H:

$$H = \sum_{i=1}^{n} s_i^2, \quad 0 < H < 1$$

Market power based on concentration can be measured by

$$L = \sum_{i=1}^{n} s_i \left(\frac{p - \mathrm{MC}_i}{p} \right)$$

which is equivalent to $L = H/\varepsilon$, where ε is the price elasticity of demand and the condition $\mathrm{MC}_i = \mathrm{MR}_i$ (marginal cost equals marginal revenue) holds in Cournot equilibrium.

A major source of increased industry concentration is the degree of scale economies associated with the size of fixed costs. For the total cost function $C = F + cx$, the minimum of average cost is c. Let the MES be the minimum scale such that average cost is $\hat{c}$. Then one obtains

$$x = \frac{F}{(\hat{\mathrm{c}} - \mathrm{c})} = \mathrm{MES}$$

where $\mathrm{AC} = \hat{c}$. Thus changes in MES are due to changes in fixed cost F. The scale economies (SE) can be measured by

$$\mathrm{SE} = \frac{\mathrm{AC}}{\mathrm{MC}} = 1 + \frac{F}{(cx)}$$

This shows that the industry concentration is greater, the greater the MES or the greater the degree of scale economies.

A second source of concentration is collusion among firms through mergers or coalitions. Brown and Chiang (2003) have analyzed the coalition structures for oligopolies and their effects on CS. In their approach we start with a set $N = \{1, 2, \ldots, n\}$ of firms with $n > 1$. There are m_j identical firms belonging to coalition j, and k is the number of coalitions. $\hat{C} = (M_1, M_2, \ldots, M_k)$ denotes the coalition structure, where M_j ($j = 1, 2, \ldots, k$) denotes the coalitions in the coalition structure. Thus for $n = 3$ five potential coalition structures can form the grand coalition $\{|1, 2, 3|\}$, denoted by $k = 1$; the three subcoalitions $\{|12|, 3\}$, $\{|13|, |2|\}$ or $\{|1|, |23|\}$, denoted by $k = 2$; and the Cournot–Nash outcome consisting entirely of singletons $\{|1|, |2|, |3|\}$, denoted by $k = 3$.

In the Cournot market game each firm sets the optimal quantity in stage 2 of the two-stage game. In the homogeneous product case the inverse demand function is

$$p = B - m_j x_j - \sum_{i \neq j}^{k} m_i x_i$$

where x_j is output for a firm in coalition j, where all firms are alike in a given coalition (i.e., symmetric). Then the profit of a typical firm in coalition j is

$$\pi_j = \left[B - m_j x_j - \sum_{i \neq j}^{k} m_i \; x_i \right] x_j - c x_j \tag{4.4}$$

On maximizing this profit one could derive the Cournot–Nash equilibrium output. If we assume that the goods produced are perfect substitutes, then one could derive the optimal output and profit as

$$x_j^* = (B - c)(m_j(k+1))^{-1}$$

$$x_j^* = m_j x_j^{*^2} = \left(\frac{1}{m_j} \right) \left[\frac{B-c}{1+k} \right]^2$$

Let $v_j = \sqrt{\pi_j}$ be the j-th firm's valuation of profit. Then

$$v_j = \frac{B-c}{1+k} \left(\frac{1}{m_j} \right)^{1/2}$$

This model can be easily generalized for the case of differentiated products. Consider an example with $n = 3$ and the inverse demand function as

$$p_i = B - x_i - a \sum_{j \neq i}^{3} x_j$$

where a is positive (negative) if the products are substitutes (complements). Profit and valuations are

$$\pi_j = \left(B - x_i - a \sum_{j \leq i}^{3} x_j - c \right) x_j \quad j = 1, 2, 3$$

$$v_i \left(\hat{C}_j \right) = \left[\pi_i \left(\hat{C}_j \right) \right]^{1/2} \tag{4.5}$$

The Cournot–Nash equilibrium outputs are then calculated by Brown and Chiang in a straightforward manner as follows:

Coalition structures	Outputs of firm $i=1,2,3$	Valuations of firm $i=1,2,3$
$\hat{C}_1=\{\|123\|\}$	$x_i=(B-c)(2(2a+1))^{-1}$	$v_i=\left(\frac{1}{2}\right)(B-c)/\sqrt{2a+1}$
$\hat{C}_2=\{\|12\|,\|3\|\}$	$x_i=\frac{(B-c)(a-2)}{2(a^2-2a-2)}$ $x_3=\frac{B-c}{-a^2+2a+2}$	$v_1=v_2=\left\{\left[\frac{1}{2}\frac{(a-2)(B-c)}{-2+a^2-2a}\right]^2(1+a)\right\}^{1/2}$ $v_3=\frac{B-c}{2-a^2+2a}$
$\hat{C}_3=\{\|1\|,\|23\|\}$	$x_i=\frac{B-c}{-a^2+2a+2}$ $x_2=\frac{(B-c)(a-2)}{2(a^2-2a-2)}$	$v_1=\frac{B-c}{2-a^2+2a}$ $v_2=v_3$, where $v_3=\left\{\left[\frac{1}{2}\frac{(a-2)(B-c)}{-2+a^2-2a}\right]^2(1+a)\right\}^{1/2}$
$\hat{C}_4=\{\|1\|,\|2\|,\|3\|\}$	$x_i=\left(\frac{1}{2}\right)\frac{B-c}{a+1}$	$v_i=\frac{B-c}{2(a+1)}$, $i=1,2,3$

Social welfare W is usually defined as the sum of PS and CS. Product surplus or profit cannot be reduced under coalition relative to the Cournot–Nash outcome since they would not form coalitions otherwise. But CS can change either way. On substituting the optimal outputs above one could compute the following:

$$W(\hat{C}_1)-W(\hat{C}_4)=\left(\frac{-3}{8}\right)\frac{(B-c)a(Ba+2B-ca-2c)}{(2a+1)(a+1)^2}\ \substack{>\\=\\<}\ 0 \text{ if } a\ \substack{<\\=\\>}\ 0$$

and

$$W(\hat{C}_3)-W(\hat{C}_4)=\frac{1}{8}\frac{(B-c)^2\,a(a^3+6a^2-6a-8)}{(a^2-2a-2)(a+1)^2}\ \substack{>\\=\\<}\ 0 \text{ if } a\ \substack{<\\=\\>}\ 0$$

This shows that if $a<0$ then $W(\hat{C}_1)>W(\hat{C}_4)$ so that the grand coalition is welfare improving. But if goods are substitutes, that is if $0<a<a_0$, where a_0 solves the equation $f(a)=a^3-2a^2-a+1=0$ then it is not welfare improving. Thus they conclude that when the grand coalition

is formed due to complementarity in goods, both PS and CS and total social welfare increase. But when goods are substitutes, that is toughness of competition, consumers suffer losses resulting from higher prices.

Two comments are in order. First, we may note from Equation (4.2) that entry n^* increases less than proportionately the market size S because of the increased pressure on profit margins when there are more firms in the industry. But it is not always true that increasing the market size leads to a decrease in concentration. For example, in the constant demand elasticity case with $p = S/X$, if costs for any firm are concave $C(x) = x^{1/r}$ with $1 < r \leq 2$, then n is bounded above by $r/(r-1)$. This means that no matter how large the market is, a finite number of firms can only survive. Secondly, we have to recognize that there are important interactions between entry and profitability. For example, Masson and Shaanan (1982) have noted that the profitability signal may reflect systematic distortions by firms eager to hide excess profits from entrants. In their model entry, $E_i(t)$ is in market shares defined as

$$E_i(t) = \theta(\pi_i(t) - \pi_i^*) \tag{4.6}$$

where the limit profits π_i^* are determined by the barriers to entry, for example advertising, absolute cost advantages and scale economies and also industry growth. Thus

$$\pi_i^* = \Sigma\, \alpha_j z_j \tag{4.7}$$

where z_j are the various entry barriers and industry growth. The relation between actual profits $\pi_i(t)$ of firm i and the target profit π_i^0 are modeled as

$$\pi_i(t) = \pi_i^0 + \beta(C(t) - 100) \tag{4.8}$$

where $C(t)$ is the industry concentration ratio. Here profits affect entry behavior via Equation (4.6) and subsequently entry feeds back profits in Equation (4.8) to the extent that concentration $C(t)$ is reduced. The critical parameters which capture this feedback are β and θ. Masson and Shaanan obtained estimates of $\theta = 0.5$ and $\beta = 0.5$ over their manufacturing data set for the United States for the years 1958–1963. Their estimates for α_j in Equation (4.7) suggest the existence of important scale but no growth effects on entry. Geroski and Masson (1987) extended this profitability model in the form

$$\pi_i(t) = \alpha_i + \beta\, \pi_i(t-1) - \lambda \sum_{\tau=0}^{\infty} E(t-\tau) \tag{4.9}$$

where β represents observed profit movements associated with the regulation of entry by incumbents and λ reflects the effects of actual entry in time lags on profits. They applied this model (Equation (4.9)) to 85 industries in the United Kingdom over 1974–1979 and the estimates of α_i, β, λ suggested rather rapid convergence of $\pi_i(t)$ to long-run profits π_i^*. They also found that both λ and β varied systematically across industries and the dynamics of entry adjustment process were slower in more advertising-intensive and highly concentrated markets. Note that the threat of entry may lead incumbent firms to remove X-efficiency and slack ridden costs we have discussed before, and hence profits after entry may be relatively constant even when entry has a dynamic effect on price. Finally, we may note that the entry by one entrant tends to reduce the potential entry gains for other potential entrants. Also, the threat of potential entry if found credible by the incumbents may lead to partial accommodation attempts by the incumbents.

4.3 Accommodation of entry

We consider in this section the Stackelberg–Spence–Dixit model, which analyzes the various aspects of the entry accommodation process. We follow the presentation of Tirole (1988) in some detail. He starts with the Stackelberg model of a duopoly, where firm 1 is the incumbent and firm 2 the new (potential) entrant. Firm 1 selects a level of capital K_1 and firm 2 K_2. Their respective profits are

$$\begin{aligned} \Pi^1(K_1, K_2) &= K_1(1-K_1-K_2) \\ \Pi^2(K_1, K_2) &= K_2(1-K_1-K_2) \end{aligned} \tag{4.10}$$

in normalized units (i.e., output may be measured in units of capital). First we assume that there is no fixed cost of entry. In a Cournot framework the optimal profit-maximizing capital of firm 2 is given by

$$K_2 = R_2(K_1) = \frac{(1-K_1)}{2}$$

where R_2 is the reaction function of firm 2. Hence firm 1 maximizes

$$\Pi^1 = K_1\left(1-K_1-\frac{1-K_1}{2}\right)$$

This yields the Cournot–Nash equilibrium

$$K_1^* = K_2^* = \frac{1}{3} \text{ with } \Pi^1 = \Pi^2 = \frac{1}{9}$$

In the Stackelberg model the leader firm (firm 1) may choose the level of capital before the follower (firm 2) and hence the leader can influence firm 2's decision. This asymmetry allows firm 1 to limit firm 2's capital level K_2. To do this it accumulates more capital than it would have done in a simultaneous equilibrium. Thus the incumbent (leader) can reduce firm 2's scale of entry. Caves and Porter (1997) call this a barrier to mobility. We may also say that firm 1 accommodates entry in that it takes entry for granted and then tries to affect firm 2's subsequent behavior.

Now introduce a fixed cost of entry f. Assume the profit function of firm 2 as

$$\Pi^2(K_1, K_2) = K_2(1 - K_1 - K_2) - f, \quad \text{if } K_2 > 0$$
$$= 0 \text{ if } K_2 = 0$$

Assume that $f < 1/16$. If firm 1 chooses $K_1 = 1/2$, firm 2 chooses $K_2 = 1/4$ with a profit of $(1/16 - f) > 0$. However, this choice of K_1 may not be optimal for firm 1, since it may be able to increase its profit by completely preventing the entry of firm 2. The capital level K_1^b that discourages entry is easily computed as

$$\max_{K_2} J = [K_2(1 - K_2 - K_1^b) - f] = 0$$

or

$$K_1^b = 1 - 2\sqrt{f} > \frac{1}{2} \tag{4.11}$$

When entry is deterred the profit of firm 1 is

$$\Pi^1 = (1 - 2\sqrt{f})[1 - (1 - 2\sqrt{f})] = 2\sqrt{f}(1 - 2\sqrt{f})$$

Clearly if f is close to 1/16, then this profit is greater than 1/8. Therefore firm 1 may be interested in completely discouraging the entry of firm 2. Firm 1 accomplishes this goal by choosing $K_1 = K_1^b$, because accumulating capital beyond K_1^b would reduce profit.

This type of model was used by Matsuyama and Itoh (1985) to analyze the desirability of protecting an infant industry, where firm 1 is a foreign firm (which has a first-mover advantage) and firm 2 a domestic one. Consider the following no protection three-stage game:

Stage 1. Firm 1 chooses capacity K_1
Stage 2. Firm 2 chooses capacity K_2, knowing K_1
Stage 3. Firms choose prices simultaneously knowing K_1 and K_2.

We may assume that the domestic firm faces an entry cost. Here it can be shown that a policy of "limited protection" which forces the foreign firm to wait until period 2 to invest domestically increases total social welfare W measured by consumers' surplus plus the profits of domestic firms.

4.4 Schumpeterian model of dynamic entry

The Schumpeterian approach to the dynamic evolution of firms in an industry has several economic features related to the entry and exit behavior of firms. Three most important features are discussed by Kamien and Schwartz (1982) and Aghion and Howitt (1992) as follows: (1) There is a positive relationship between innovation and monopoly, where the latter may yield excess or above-normal profits; (2) large firms in an industry are more than proportionately innovative than small firms; and (3) streams of innovations in different forms, for example R&D, new technologies, new markets and new products make old technologies or products obsolete. This is called "creative destruction", which generates positive externalities for future research and innovations, although they exert a negative externality effect on the incumbent firms.

As Kamien and Schwartz have pointed out, the monopoly power generated by the innovation process has two important sources: the ability to retard imitation of the innovation and also to internalize the benefits of innovation through internal financing and extending the primary innovation to other products.

The major advantage of a large firm over a small one lies in its superior ability to utilize the complementary effects of its R&D skills, which have scale economies due to indivisibilities. Note that the incentive to innovate for the industry with the more elastic demand curve is more for the large firms and hence any innovation-induced price reduction yields a greater expansion in output for the more elastic market. The large firms here enjoy superior advantage over small firms due to their superior abilities to adopt cost-reducing innovations over time.

Lukach *et al.* (2005) have developed a dynamic duopoly model in Cournot–Nash framework where one firm is bigger in the sense that its marginal cost of production is lower. This model uses a linear inverse demand curve in normalized units as

$$p_t = 1 - X_t, \quad X_t = \text{total output} \tag{4.12}$$

with cost comprising two components and production costs and R&D costs of firm i as

$$C_i = (a_i - k_{it} - \beta k_{jt})x_{it}; \quad i,j = 1,2; \quad i \neq j \tag{4.13}$$

$$C_i^R = g_i \frac{z_{it}^2}{2}, \quad i = 1,2 \tag{4.14}$$

where z_{it} is the R&D effort and k_{it} represents the accumulated knowledge capital through R&D of firm i. The dynamic evolution of knowledge capital is of the form

$$k_{it} = (1-\delta)k_{it-1} + z_{it} \tag{4.15}$$

Assume that firm 1 is bigger and has advantages over both research and product costs over firm 2, that is $a_1 < a_2$ and $g_1 < g_2$.

Consider a two-stage game. In the second stage, each firm maximizes its total profit function as

$$\underset{\{x_{it}\}}{\text{Max}}\ J = \sum_{t=0}^{T-1} (1+r)^{-t} \left[(1-X_t)\, x_{it} - (a_i - k_{it} - \beta k_{jt})x_{it} - g_i \frac{z_{it}^2}{2} \right]$$

subject to Equations (4.12)–(4.15). The optimal solution can be easily computed as

$$x_{it}^* = \frac{1}{3}\,(1 - 2a_1 + a_j + (2-\beta)\, k_{it} + (2\beta - 1)\, k_{jt}) \quad i,j = 1,2; \ i \neq j$$

In the first stage, each firm chooses its R&D variables optimally as

$$\underset{\{z_{it}\}}{\text{Max}}\ \pi_i^* = \sum_{t=0}^{T-1} (1+r)^{-t} \left[\left(\frac{1}{9}\right) (1 - 2a_i + a_j + (2-\beta)\, k_{it} + (2\beta-1)\, k_{jt})^2 - g_i \frac{z_{it}^2}{2} \right]$$

subject to Equations (4.12)–(4.15).

Since this is a standard linear quadratic control problem, its optimal solution, if it exists, can be written as a feedback rule:

$$z_t^* = H_t k_t + h_t \tag{4.16}$$

with

$$z_t = \begin{pmatrix} z_{1t} \\ z_{2t} \end{pmatrix}, \quad k_t \begin{pmatrix} k_{1t} \\ k_{2t} \end{pmatrix}$$

H_t, h_t depend on all the parameters. Several implications of the optimal decision rule and the R&D investment trajectories are illustrated both theoretically and numerically by simulation. First of all, the more efficient firm (i.e., the bigger firm with lower marginal cost) invests more in R&D and produces more than the less efficient one. In the beginning of the planning horizon the R&D investment for each firm rises over time but at a decreasing rate. The behavior of optimal output and knowledge capital exhibit a similar pattern of behavior. Clearly the bigger firm with more ability to internalize the spillover effects tends to invest higher in R&D and knowledge capital than the smaller firm. Thus in case of $n > 2$ the higher the industry concentration, the greater may be the incentive to invest in R&D and knowledge capital than a fully competitive framework.

Now we consider the Schumpeterian model of entry, where the research sector is not fully competitive. Aghion and Howitt (1992, 1998) have discussed this model in some detail in terms of barriers to entry in R&D set up by the monopolistic firms as entry deterrence. This type of model may be appropriate for the problem of foreign market entry by a firm based in a home country, which has been discussed by Buckley (2000).

In a competitive market the R&D investment may be made by any firm, whereas in an oligopolistic setting it is done by the incumbent firms The level of research (i.e., amount of effort or labor allocated to research) w_t is written as

$$w_t = \lambda(V_{t+1} - V_t)$$

with

$$rV_t = \pi_t + \lambda n_t(V_{t+1} - V_t) \tag{4.17}$$

Here V_{t+1} is the discounted expected payoff to the $(t+1)$-th innovation determined by the arbitrage asset condition

$$rV_{t+1} = \pi_{t+1} - \lambda n_{t+1}\ V_{t+1} \tag{4.18}$$

which says that the expected income generated by a license on the $(t+1)$-th innovation during a unit time interval is equal to the profit flow (π_t) attainable by the $(t+1)$ intermediate good monopolist minus the expected capital loss that will occur when the $(t+1)$-th innovator is replaced by a new innovator resulting in losses V_{t+1}. The flow probability of this loss is the arrival rate λn_{t+1}, where n_t is the amount of labor used in research. In other words, the right-hand side of Equation (4.18) expresses the fact that the incumbent innovator internalizes the positive intertemporal spillover externality of the current innovation on future research.

In this model if we introduce an entry fee to enter the R&D network in the market, then it can be shown that a reduction in this fee would increase the arrival rate of innovations per research worker. Note that the ability to internalize the effects of R&D by the incumbent large firms provides the major incentive to invest in the future stream of innovations in the Schumpeterian framework.

Consider an example of the problem of foreign market entry of a firm based in a home country which is seeking to sell in a foreign market. Foreign market demand for the product is infinitely elastic at a price p. The firm may enter the foreign market by either owning or controlling production activity or distribution activity. Buckley then compares the relative costs and profits of various market entry strategies such as (1) where production is located and (2) whether ownership is outright or shared through international joint ventures. The final choice depends on the maximum expected profits. Note that the relative costs of different strategies depend on the information costs and transaction costs. When these costs are higher, entry would be discouraged as in the Schumpeterian model, where increasing the entry fee is introduced in the innovations model.

Subsidy to R&D investments may have very similar effects as the reduction in entry fee in the Schumpeterian model. Thus Brander and Spence (1983) developed a model of R&D investment in a Cournot duopoly, where it is shown that a rise in home R&D subsidy raises the home R&D level but reduces the foreign R&D level. Ishii (2000) has generalized this model to include demand uncertainty faced by each firm. This model assumes the following demand and unit cost functions for two firms (one home firm and one foreign firm) with identical products exported abroad:

$$
\begin{aligned}
&p = a - b(x_1 + x_2), \quad a, b > 0 \\
&c_i = c_i(x_i,\ z_i) = g_i(z_i)\ x_i^2 \quad \text{with } g_i'(z_i) < 0,\ g_i''(z_i) > 0
\end{aligned}
\tag{4.19}
$$

where z_i is cost-reducing R&D for firm i and prime denotes derivatives. Total cost C_i is

$$
C_i = g_i(z_i)\ x_i^2 + v_i z_i \tag{4.20}
$$

The state subsidy for R&D is

$$S_i(z_i) = f_i + s_i z_i, \;\; F_i = \text{fixed component}$$

The profit function is

$$\pi_i = [a - b(x_1 + x_2)]\, x_i - g_i(z_i)\, x_i^2 - v_i z_i - S_i(z_i) \tag{4.21}$$

On using this profit function the optimal outputs of this Cournot model are

$$\begin{aligned} x_1 &= \frac{a[b + 2g_2(z_2)]}{D} \\ x_2 &= \frac{a[b + 2g_1(z_1)]}{D} \end{aligned} \tag{4.22}$$

where

$$D = 3b^2 + 4bg_1(z_1) + 4bg_2(z_2) + 4g_1(z_1)\, g_2(z_2)$$

This solution (Equation (4.22)) is derived in two stages. In the first stage both firms face demand uncertainty since the parameter a of the demand function in Equation (4.19) is unknown to them. In this stage firm i chooses its R&D level z_i given the rival's z_j so as to maximize its own expected utility of profits $E_i[u(\pi_i)]$. In the second stage the firms decide their optimal outputs when the demand uncertainty is resolved and optimal levels of R&D have already been chosen in the first stage. Thus the firm's reaction functions under demand uncertainty can be derived as

$$\begin{aligned} M_i = \frac{\partial E_i[u_i(\pi_i)]}{\partial z_i} &= E_i\left[\left(\frac{\partial \pi_i}{\partial z_i}\right) u_i'(\pi_i)\right] \\ &= E_i\left[\left\{ bx_i \left(\frac{\partial x_j}{\partial z_j}\right) - g_i'(z_i)\, x_i^2 + v_i - s_i \right\} u_i'(\pi_i)\right] = 0 \end{aligned} \tag{4.23}$$

To obtain an interior solution it is assumed that $s_i < v_i, \; i = 1, 2$. Now consider the effects of a change in home-fixed subsidy F on the optimal R&D levels. When there is no demand uncertainty, F_1 has no effect on firm's optimal R&D levels since $\partial F_1/\partial z_1 = 0$. But with demand uncertainty, F_1 is included in the first-order condition of a home firm, where a change in F_1 shifts a home R&D reaction curve while it does not shift a foreign R&D reaction curve. Thus when an absolute risk aversion function $R_A(\pi_1)$ of a home firm is decreasing (increasing), a rise (fall) in the home-fixed R&D subsidy raises (reduces) the home R&D level. Ishii (2000) has also shown that the effect of a change in the home marginal R&D subsidy s_1 on the home R&D level is positive when $R_A(\pi_1)$ is decreasing. Thus it is clear that it is more difficult for the home government to decide on an optimal R&D policy when significant demand uncertainty is present.

Instead of the subsidy problem faced by the government, Ponssard (1979) has used the uncertainty of knowing the parameter a in the inverse demand function (Equation (4.19)) to construct an information game for n oligopoly firms. There are two groups of firms: $k = n_1$ are informed and $n_2 = n - n_1$ are uninformed about the true value of the demand shift parameter a. Then he has shown that the expected profits at the Cournot–Nash equilibrium point for the informed firms are increased by var $a/[b(k+1)^2]$, where $k = I_1$ (the number of informed firms). If all firms are informed, that is $k = I$, then the expected consumers' surplus is increased by

$$\Delta CS = k^2 \text{ var} \frac{a}{[2b(k+1)]^2} \tag{4.24}$$

Now if the probability distribution $f(a)$ of a is such that the variance increases with the expected value, that is $\partial\sigma_a^2/\partial\bar{a} > 0$, where $\bar{a}$ and σ_a^2 are the means and variances, then the increase in CS (Equation (4.24)) is positively proportional to the expected value $\bar{a}$, that is the higher $\bar{a}$ implies a larger increase in CS.

Thus when the monopolistic firms can absorb all the increased consumers' surplus through internalizing, they have the additional incentive to innovate in Schumpeterian theory. Note also that their total reward increases with the size of the market.

In the context of global markets for technology-intensive products today entry has to be viewed in terms of evolution of new and emerging markets. Geroski (2003) has argued that supply push innovation of the Schumpeterian variety is no longer concentrated on a single good or service. Rather they expand the market in a variety of forms. Entry rates in very young markets are extremely high. This is mainly due to low barriers to entry, that is the ease of introducing new technological possibilities through new innovation.

There is another aspect of the new variety of technological competition which is driving the new markets today and its expansion in a global framework. In the early phase of this competitive struggle, competition is basically between different product designs. As Geroski (2003) has pointed out, the survival of particular firms at this stage depends very critically on the viability of their design and their core competence in innovation efficiency.

In the second stage when a dominant design is established, the competition between designs is no longer an issue and this is replaced by competition within designs. This second-stage competition sparks a race down the learning curves, triggering large R&D investments in order to capture the economies of scale and scope.

4.5 An empirical application to high-tech industries

We consider in this section two empirical applications in the modern computer and pharmaceutical industries. The application focuses on three basic features: (1) we estimate the number of efficient firms by means of the method of DEA and test if the efficient firms have grown over time much more than the inefficient ones; (2) to what extent the R&D inputs have contributed to the efficiency gains over time and (3) the supply function elasticity for different firms in the industry.

The efficiency estimate is based on the DEA method, which estimates the Pareto efficiency frontier through a sequence of linear programming models. This method is discussed in some detail by Sengupta (2003).

The supply function estimates may be considered as a surrogate view of the entry behavior. This has been the method followed by Veloce and Zellner (1985), who applied it to estimate the entry equations for the Canadian household furniture industries as discussed before. The evolution of high-tech industries in modern times has been profoundly affected by innovations in different forms such as new product designs and new software developments. Research and development spending captures the key elements of the dynamic innovation process. Several features of R&D investment by firms are important in the dynamic evolution of an industry. First of all, R&D spending not only generates new knowledge about technical processes and products but also enhances the firm's capability to improve the stock of existing "knowledge capital". This is the process of learning that has cumulative impact on industry growth. Secondly, growth of R&D spending helps in expanding the growth of sales or demand through new product variety and quality improvements. This has often been called economies of scale in demand in the modern theory of hyper-competition analyzed by Sengupta (2004). Thirdly, the R&D investment within a firm has a spillover effect in the industry as a whole. This is because R&D spending yields externalities in the sense that knowledge acquired by one firm spills over to other firms and very often knowledge spread in this way finds new applications both locally and globally and thereby stimulates further innovative activity and R&D intensity in other firms.

Our objective here is twofold. One is to incorporate R&D investment into the DEA efficiency models and thereby show its impact on market demand and efficiency. Secondly, we apply these efficiency models in two modern industries: computers and pharmaceuticals. These empirical applications apply a two-stage approach to economic efficiency. In the first stage the efficient levels of R&D inputs are determined for the DEA efficiency firms and in the second stage we estimate by a regression model the role of R&D spending in total sales. This type of analysis is especially important for the pharmaceutical industry, since the share of R&D spending in total cost is much higher here, as the development of new medicines requires substantial spending on research.

4.5.1 Efficiency models in R&D

Three types of R&D models are developed here for empirical and theoretical applications. One emphasizes the cost-reducing impact of R&D inputs. This may be related to the learning by doing implications of knowledge capital. Secondly, the impact on output growth through increases in R&D spending is formalized through a growth efficiency model. Here a distinction is drawn between *level* and *growth* efficiency, where the former specifies a static production frontier and the latter a dynamic frontier. Finally, the market structure implications of R&D spending are analyzed in a Cournot-type industry, where R&D spending is used as a marketing strategy just like advertisement.

Denote average cost by c_j/y_j, where total cost c_j excludes R&D costs denoted by r_j. Then we set up the DEA model with radial efficiency scores θ.

$$
\begin{aligned}
&\min \theta \\
&\text{subject to } \sum_{j=1}^{n} c_j\lambda_j \le \theta c_h, \quad \sum_j r_j\lambda_j \le r_h \\
&\qquad \sum_j r_j^2\lambda_j = r_h^2, \quad \sum y_j\lambda_j \ge y_h \qquad\qquad (4.25)\\
&\qquad \sum \lambda_j = 1, \quad \lambda_j \ge 0, \quad j \in I_n = \{1, 2, \ldots, n)
\end{aligned}
$$

By using dual variables $\beta_1, \beta_2, \beta_3, \alpha, \beta_0$ and solving the LP model (4.1), for an efficient firm j we obtain $\theta^* = 1.0$, zero for all slack variables and the following average cost frontier

$$c_j^* = \beta_0^* - \beta_2^* r_j + \beta_3^* r_j^2 + \alpha^* y_j \qquad (4.26)$$

since $\beta_1^* = 1.0$ if $\theta^* > 0$. Thus if R&D spending r_j rises, average cost c_j falls if $2\beta_3^*\ r_j < \beta_2^*$. If we replace r_j by cumulative R&D knowledge capital R_j as in learning-by-doing model, where R_j is cumulative experience, then the AC frontier (Equation (4.26)) becomes

$$c_j^* = \beta_0^* - \beta_2^*\ R_j + \beta_3\ {}^*R_j^2 + \alpha^*\ y_j \tag{4.27}$$

So long as the coefficient β_3^* is positive, r_j may also be optimally chosen as r^* if we extend the objective function in Equation (4.25) as $\min \theta + r$ and replace r_h by r. In this case we obtain the optimal value of R&D spending r^* as

$$r^* = (2\beta_3^*)^{-1}\ (1 + \beta_2^*) \tag{4.28}$$

A similar result follows when we use the cumulative R&D spending R_j or R.

Two simple extensions of the cost frontier model (4.25) can be derived. The first is to extend the case to multiple outputs and multiple R&D inputs. We have to replace single output y_j to y_{kj}, $k \in I_m$, with m research inputs. Then the AC frontier (Equation (4.26)) would appear as

$$c_j^* = \beta_0^* - \sum_{i=1}^{m} \beta_{zi}^* r_{ij} + \sum_{i=1}^{m} \beta_{3i} r_{ij}^2 + \sum_{k=1}^{s} \alpha_k^*\ y_{kj}$$

Secondly, we may formulate the model in terms of total costs rather than average costs as

$$\begin{aligned}
&\min\ \theta \\
&\text{subject to}\ \sum c_j \lambda_j \le \theta\ c_h, \quad \sum_j r_j \lambda_j \le r_h \\
&\qquad \sum_j r_j^2 \lambda_j = r_h^2, \quad \sum_j y_j \lambda_j \ge y_h, \quad \Sigma y_j^2 \lambda_j \ge y_h^2 \\
&\qquad \sum \lambda_h = 1, \quad \lambda \ge 0
\end{aligned}$$

In this case the total cost frontier becomes

$$c_j^* = \beta_0^* - \beta_2^*\ r_j + \beta_3^*\ r_j^2 + \alpha_1^* y_j + \alpha_2^* y_j^2$$

where α_1^*, $\alpha_2^* \geq 0$. If the intercept term β_0^* is positive, then average cost for the j-th efficient unit can be reduced to

$$\mathrm{AC}_j = \left(\frac{\beta_0^*}{y_j}\right) + \alpha_1^* + \alpha_2^* y_j + \frac{\left(\beta_3^* r_j^2 - \beta_2^* r_j\right)}{y_j}$$

On setting its derivative to zero we obtain the optimal level of output y_j^* for fixed levels of r_j as

$$y_j^* = \left[\frac{\left(\beta_0^* + \beta_3^*\, r_j^2 - r_j\right)}{\alpha_2^*}\right]^{1/2} \tag{4.29}$$

If research costs r_j are already included in total costs, then the optimal level of efficient output y_j^* in (4.29) reduces to

$$y_j^* = \left(\frac{\beta_0^*}{\alpha_2}\right)^{\frac{1}{2}} \tag{4.30}$$

The associated value of minimum AC then becomes

$$\mathrm{AC}_{\min} = \alpha_1^* + 2\sqrt{\beta_0^* \alpha_2^*}$$

This level of cost $\mathrm{AC}_{\min}$ may be used to define MES of efficient firm j. Note that this measure is more comprehensive and structural than the most traditional productive scale size (MPSS) used in DEA models.

Now consider a second type of model where growth efficiency is considered. Several types of models of growth efficiency frontier and their comparison with level efficiency have been discussed by Sengupta (2003). Here we consider a firm j producing a single composite output y_j with m inputs x_{ij} by means of a log-linear production function

$$y_j = \beta_0 \prod_{i=1}^{m} e^{B_i} x_{ij}^{\beta_i}; \quad j = 1, 2, \ldots, N \tag{4.31}$$

where the term e^{B_i} represents the industry effect or a proxy for the share in total industry R&D. On taking logs and time derivatives of both sides of (4.31) one can then easily derive the production frontier

$$Y_j \leq \sum_{i=0}^{m} b_i X_{ij} + \sum_{i=1}^{m} \phi_i \hat{X}_i \tag{4.32}$$

where

$$b_i = \beta_i, b_0 = \frac{\dot{\beta}_0}{\beta_0, X_{0j}} = 1, \quad j = 1, 2, \ldots, N$$

$$e^{B_i} = \phi_i \hat{X}_i, \quad X_{ij} = \frac{\dot{x}_{ij}}{x_{ij}}, \quad Y_j = \frac{\dot{y}_j}{y_j}, \quad \hat{X}_i = \frac{\sum_{j=1}^{N} \dot{x}_{ij}}{\sum_{j=1}^{N} x_{ij}}$$

Dot denotes time derivative.

Note that b_0 denotes here technical progress representing innovation efficiency or productivity growth (i.e., Solow-residual) and ϕ_i denotes the input specific industry efficiency parameter. Sengupta (2004) has applied models of joint research ventures with other firms in the industry which attempt to increase access efficiency in investments on research and development.

We now consider how the following LP model is solved to test the relative efficiency of each firm k in an industry of n firms:

$$\begin{aligned} &\text{Min } C_k = \sum_{i=0}^{m} (b_j X_{ik} + \phi_i \hat{X}_i) \\ &\text{subject to } \sum_{i=0}^{m} (b_i X_{ij} + \phi_i \hat{X}_i) \geq Y_j, \quad j = 1, 2, \ldots, n \\ &\qquad b_0 \text{ free in sign, } b_1, b_2, \ldots, b_m \geq 0, \phi_i \geq 0 \end{aligned} \tag{4.33}$$

Let b^* and ϕ^* be the optimal solutions for the observed input–output data set X_{ij}, $\hat{X}_i$ and y_j, $j = 1, 2, \ldots, n$ with all slack variables zero. Then firm k is growth efficient if

$$Y_k = b_0^* + \sum_{i=1}^{m} \left(b_i^* X_{ik} + \phi_i^* \hat{X}_i \right) \tag{4.34}$$

If, however, we have

$$b_0^* + \sum_{i=1}^{m} \left(b_i^* X_{ik} + \phi_i^* \hat{X}_i \right) > Y_k \tag{4.35}$$

then the k-th firm is not growth efficient, since the observed output Y_k is less than the optimal output $Y_k^* = b_0^* + \sum_{i=1}^{m} (b_i^* X_{ik} + \phi_i^* \hat{X}_i)$. Note that this nonparametric method has several flexible features. First of all, on varying k over $1, 2, \ldots, n$ one could group the firms into two subsets, one efficient thus satisfying Equation (4.34) and the other inefficient satisfying Equation (4.35). Secondly, if the input–output data set over time is available, one could estimate the parameters $b_0^*(t)$, $\phi_j^*(t)$ and $b_i^*(t)$ for all $t = 1, 2, \ldots, T$. The output efficiency scores $\varepsilon_k^*(t) = Y_k(t)/Y_k^*(t)$ can also be computed for the efficient and inefficient units. Thirdly, if the innovation efficiency is not input-specific, that is $e^{B_i} = \phi t$, then one could combine the two measures of dynamic efficiency as $b_0^* + \phi^* = \tilde{b}_0^*$, say, representing innovation and access efficiency. In this case the dual problem for Equation (4.6) can be simply formulated as

$$\begin{aligned} &\text{Max } \mu, \text{ subject to } \sum_{j=1}^{N} X_{ij}\lambda_j \leq X_{ik}, \quad i = 0, 1, 2, \ldots, m \\ &\sum_{j=1}^{N} Y_j\lambda_j \geq \mu Y_k; \quad \sum_{j=1}^{N} \lambda_j = 1, \quad \lambda_j \geq 0 \end{aligned} \tag{4.36}$$

An input-based efficiency model can be similarly specified as

$$\begin{aligned} &\text{Min } \theta, \text{ subject to } \sum_{j=1}^{N} X_{ij}\lambda_j \leq \theta X_{ik}, \quad i = 0, 1, \ldots, m \\ &\sum_{j=1}^{N} Y_j\lambda_j \geq Y_k; \qquad \Sigma\lambda_j = 1, \quad \lambda_j \geq 0 \end{aligned} \tag{4.37}$$

If the optimal values μ^* and θ^* are unity, then the unit k is growth efficient, otherwise it is inefficient. As before the efficiency scores $\mu^*(t)$, $\theta^*(t)$ can be computed over time if the time series data on inputs and outputs for each firm are available. Since some of the inputs are services of capital inputs, their impact on supply side economies of scale can be captured by the sum of the respective coefficients of production.

Finally, we note that the growth efficiency models (Equations (4.33)–(4.37)) can be compared with the static model for testing the level efficiency of firm k. For instance the models analogous to Equations (4.33) and (4.37) would appear as follows:

$$\begin{aligned} &\text{Min } C_k = \tilde{\beta}_0 + \sum_{i=1}^{m} \left(\tilde{\beta}_i \ln x_{ik} + \tilde{\phi}_i \ln x_i\right) \quad \text{where } x_i = \sum_{j=1}^{N} x_{ij} \\ &\text{subject to } \quad \tilde{\beta}_0 + \sum_{i=1}^{m} \left(\tilde{\beta}_i \ln x_{ij} + \tilde{\phi}_i \ln x_i\right) \geq y_j \\ &\tilde{\beta}_0 \text{ free in sign, } \tilde{\beta}_i, \tilde{\phi}_i \geq 0 \quad j = 1, 2, \ldots N \end{aligned} \tag{4.38}$$

and

$$\text{Min } \tilde{\phi}, \text{ subject to } \sum_{j=1}^{N} x_{ij}\tilde{\lambda}_j \leq \tilde{\theta} x_{ik}$$
$$\sum_{j=1}^{N} y_j\tilde{\lambda}_j \geq y_k; \quad \sum \tilde{\lambda}_j = 1, \quad \tilde{\lambda}_j \geq 0. \tag{4.39}$$

The time series values of efficiency scores $\tilde{\theta}^*(t)$ of level efficiency may then be compared with those $\theta^*(t)$ of growth efficiency. If innovation and access efficiency by R&D spending are most dominant characteristics of firms on the leading edge of growth frontier, this would be captured more strongly by the dynamic efficiency scores $\theta^*(t)$ and their trend over time.

We now consider an empirical application to the computer industry based on Standard and Poor's Compustat data, where on economic grounds a set of 40 firms (companies) in the computer industry over a 16-year period 1984–1999 are selected by way of illustrating the concepts of dynamic efficiency analyzed before. The companies included here comprise such well-known firms as Apple, Compaq, Dell, IBM, HP and also lesser-known firms such as AST Research, Pyramid Tech, Toshiba, NBI and Commodore, and so on. Due to a variety of differentiated products, a composite output represented by total sales revenue is used as the single output (y_j) for each company. Ten inputs are selected from the Compustat Database representing both financially related input variables such as manufacturing costs and marketing costs and also "net capital employed" at the end of the reporting period representing input variables such as working capital, plant and equipment and other fixed assets. Also we use a proxy variable (x_{10}) for all nondiscretionary inputs represented by advertising expenditures by the competing firms. Three inputs in manufacturing costs are x_1 for raw material costs, x_2 for direct labor and x_3 for overhead expenses. Three inputs for marketing costs include x_4 for advertising, x_5 for R&D expenses and x_6 for other selling and administrative expenses. Net capital employed in dollars includes x_7 for working capital, x_8 for net plant and equipment and x_9 for other fixed assets. Finally, x_{10} represents a proxy variable for competitive pressure exerted by the competitors to a given firm j. Thus we have used empirical data of 40 firms each producing one output (y_j) with 10 inputs (x_{ij}; $i = 1, 2, \ldots, 10$ and $j = 1, 2, \ldots, 40$).

Three types of empirical applications are discussed here. The first characterizes the two subsets of efficient (N_1) and inefficient (N_2) firms, where $N = N_1 + N_2 = 40$. Since efficiency varies over time we consider the median efficiency level $\bar{\varepsilon}^*$ over the period 1984–1998 and N_1 includes all firms with efficiency level ε_k^* higher than $\bar{\varepsilon}^*$, where $\varepsilon_k^*(t)$ is defined by the LP model (Equation (4.33)) and likewise for the level efficiency score $\tilde{\theta}^*(t)$ when we apply the LP model given in Equation (4.38). One point stands out most clearly in the estimates of Table 4.1. Dynamic efficiency in the form of technical progress and R&D efficiency explain the major share of growth efficiency of the efficient firms. Since these two sources of efficiency are good proxy variables for innovation and access efficiency, it is clear that hypercompetition accentuates the divergence of less efficient firms from the cutting-edge growth frontier. The market pressure coefficient (b_{10}) is also very important.

Secondly Table 4.2 shows the output growth of efficient and inefficient firms. The growth efficient firms exhibit much faster growth than the inefficient firms. Furthermore the inefficient firms exhibit a logistic trend with the rate of growth declining at a slow rate. The latter aspect may reflect a tendency to exit from the industry. Finally, Table 4.3 compares the two types of efficiency: the level efficiency and growth efficiency. The efficient firms reveal a much stronger showing in terms of growth efficiency than level efficiency. This implies that in the

Table 4.1 Sources of growth efficiency

	Technical progress (%) b_0	**R&D efficiency (%) b_5**	**Plant and equipment efficiency b_8**	**Market pressure b_{10}**
Efficient firms ($N_1 = 12$)	35	39	21	21
Inefficient firms ($N_2 = 28$)	12	13	18	19

Table 4.2 Output trends over time ($\Delta y(t) = a_0 + a_1 y(t)$)

	a_0	a_1	a_2	R^2
Efficient firms ($N_1 = 12$)	−0.602	0.019*	–	0.961
Inefficient firms ($N_2 = 28$)	–	0.009*	−0.004	0.954

* denotes significant t at 5 percent level and a_2 is the coefficient for a logistic trend.

Table 4.3 Level efficiency versus growth efficiency

	Median score		Mean deviation		Coefficient of variation	
	$\overline{\theta}^*$	$\overline{\varepsilon}^*$	θ^*	ε^*	θ^*	ε^*
Efficient firms	0.951	0.982	0.105	0.043	0.457	0.231
Inefficient firms	0.895	0.891	0.101	0.014	0.356	0.247

Efficient firms: $\tilde{\theta}_t^* = 0.013 + 0.957^{**}\tilde{\theta}_{t-1}^*$
Inefficient firms: $\tilde{\theta}_t^* = 0.028 + 0.867^{**}\tilde{\theta}_{t-1}^*$
Efficient firms: $\varepsilon^*(t) = 0.003 + 0.978^{**}\varepsilon^*(t-1)$
Inefficient firms: $\varepsilon^*(t) = 0.012 + 0.879^{**}\varepsilon^*(t)$
θ_t^* = level efficiency score; $\varepsilon^*(t)$ = growth efficiency score.
** denotes significant t-values at 1 percent level.

computer industry it is more relevant to apply a dynamic production frontier involving the growth of various inputs and output.

Finally, we consider the market structure implications of R&D spending by firms in competitive and Cournot-type markets.

4.5.2 Efficiency in computer industry

Recent times have seen intense competition and growth in high-tech industries such as semiconductors, microelectronics and computers. Product and process innovations, economies of scale and learning by doing have intensified the competitive pressure leading to declining unit costs and prices. Sengupta (2003, 2004) has analyzed in some detail the growth trends in computer industry by using DEA models and regression estimates based on the efficient estimates of the DEA models.

To consider the dynamic impact of R&D investments we have used growth efficiency models in DEA framework, one based on a dynamic production frontier, the other on a dynamic cost frontier. We have used Standard and Poor's Compustat database with SIC codes 3570 or 3571 as described before, over the period 1985–2000, covering 40 firms. The dynamic production frontier model uses a nonradial efficiency score $\theta_i(t)$ specific to input i as follows.

$$\min \sum_{i=1}^{m} \theta_i(t)$$

$$\text{subject to} \quad \sum_{j=1}^{n} \tilde{x}_{ij}(t)\lambda_j(t) \leq \theta_i(t)\, \tilde{x}_{ih}(t), \quad i \in I_m$$

$$\sum_{j=1}^{n} \tilde{y}_j(t)\lambda_j(t) \geq \tilde{y}_h\ (t) \tag{4.40}$$

$$\sum_j \lambda_j(t) = 1, \quad j \in I_n; \quad t = 1, 2, \ldots, T$$

Here $\tilde{z}_j(t) = \Delta z_j(t)/z_j(t)$, $z_j(t) = x_{ij}(t)$, $y_{(j)}(t)$ denote percentage growth. For an efficient firm j on the dynamic production frontier we would have

$$\frac{\Delta y_j(t)}{y_j(t)} = \beta_0^* + \sum_{i=1}^{m} \beta_i^* \frac{(\Delta x_{ij})}{x_{ij}(t)} \tag{4.41}$$

where β_0^* is free in sign and β_i^* values are nonnegative. Since one could derive the above model from a log-linear or Cobb–Douglas production function, one could measure the scale $S = \sum_{i=1}^{m} \beta_i^*$ by the sum of input coefficients, and β_0^* in Solow-type growth models measures technological progress if it is positive (and vice versa). Thus by using a four-year moving average one could obtain long-run changes in scale $S(\tau)$ and technological progress $\beta_0^*(\tau)$, where τ may denote, for example, a three-year moving average. Thus if $\beta_0^*(3) > \beta_0^*(2) > \beta_0^*(1) > 0$ then the technology is improving and likewise $S(3) > S(2) > S(1) > 0$ for scale improvement.

A cost-oriented version of the model given in Equation (4.40) may be written as

$$\min\ \theta(t)$$

$$\text{subject to} \quad \sum_{j=1}^{n} \tilde{C}_j(t)\ \mu_j(t) \leq \phi(t)\,\tilde{C}_h(t)$$

$$\sum_j \tilde{y}_j(t)\ \mu_j(t) \geq \tilde{y}_h\ (t) \tag{4.42}$$

$$\Sigma\ \tilde{y}_j^2\ \mu_j(t) = \tilde{y}_h^2(t)$$

$$\Sigma\ \mu_j = 1, \quad \mu_j \geq 0, \quad j \in I_n$$

where we have used total cost and total output by $C_j(t)$ and $y_j(t)$ and the quadratic output constraint is written as an equality so that the nonlinearity effect would make the cost frontier strictly convex. The dynamic cost frontier for an efficient firm j can then be written as

$$\tilde{C}_j(t) = \gamma_0^* + \gamma_1^* y_j(t) + \gamma_2^* y_j^2(t) \tag{4.43}$$

If one excludes R&D spending from total costs C_j and denotes it by $R_j(t)$, then the dynamic cost frontier can be specified as

$$\frac{\Delta C_j(t)}{C_j(t)} = \beta_0^* + \beta_1^* \left(\frac{\Delta y_j(t)}{y_j(t)} \right) - \beta_2^* \left(\frac{\Delta R_j}{R_j} \right) \tag{4.44}$$

Here β_1^*, β_2^* are nonnegative and β_0^* is free in sign. Here the elasticity coefficient β_2^* estimates in the DEA framework the influence of the growth of R&D spending on the growth of costs.

Table 4.4 estimates the nonradial efficiency measures specified in Equation (4.40), where all inputs are grouped into three inputs: R&D, net plant and capital expenditure, and cost of goods sold excluding R&D spending. These are denoted by x_1, x_2 and x_3 respectively. The importance of the R&D input is clearly revealed by its efficiency score. Companies which have experienced substantial growth in sales have also exhibited strong efficiency in R&D input utilization, for example Dell, Sequent, Sun Microsystems and Data General.

Table 4.5 shows the impact of R&D inputs on growth efficiency through the cost frontier model shown in Equation (4.44). Note that the R&D spending defined here include not only software development and research but also all types of marketing and networking expenses. Data limitations prevent us from considering only the research-based expenses here. The companies which are leaders in growth efficiency

Table 4.4 Nonradial average efficiency measures $\theta_i^*(t)$ based on the growth efficiency LP model

	1985–1989			1990–1994			1995–2002		
	$\theta_1^*(t)$	$\theta_2^*(t)$	$\theta_3^*(t)$	$\theta_1^*(t)$	$\theta_2^*(t)$	$\theta_3^*(t)$	$\theta_1^*(t)$	$\theta_2^*(t)$	$\theta_3^*(t)$
Dell	0.61	0.44	0.47	1.0	1.0	1.0	1.0	1.0	1.0
Compaq	0.40	0.54	0.50	1.0	1.0	1.0	0.33	0.60	0.75
HP	0.49	1.0	0.47	0.55	0.80	1.0	1.0	1.0	1.0
Sun	1.0	1.0	1.0	1.0	1.0	1.0	0.42	0.24	0.67
Toshiba	0.49	0.62	0.72	1.0	1.0	1.0	1.0	1.0	1.0
Silicon Graphics	1.0	1.0	1.0	1.0	1.0	1.0	0.25	0.25	0.38
Sequent	1.0	1.0	1.0	1.0	1.0	1.0	0.50	0.54	0.48
Hitachi	0.40	0.68	0.65	1.0	1.0	1.0	0.94	0.84	1.0
Apple	0.52	0.69	0.64	0.51	0.44	0.76	1.0	1.0	1.0
Data General	1.0	1.0	1.0	1.0	1.0	1.0	0.48	0.54	0.77

Three inputs are x_1 = R&D expenditure, x_2 = net plant and equipment expenditure and x_3 = cost of goods sold. $\theta_i^*(t)$ corresponds to x_i for $i = 1, 2, 3$.

Table 4.5 Impact of R&D inputs on growth efficiency based on the cost-oriented model

Company	1985–1989		1990–1994		1995–2000	
	θ^*	β_2^*	θ^*	β_2^*	θ^*	β_2^*
Dell	1.00	2.71	1.00	0.15	0.75	0.08
Compaq	0.97	0.03	1.00	0.002	0.95	0.001
HP	1.00	1.89	0.93	0.10	0.88	0.002
Sun	1.00	0.001	1.00	0.13	0.97	1.79
Toshiba	0.93	1.56	1.00	0.13	0.97	1.79
Silicon Graphics	0.99	0.02	0.95	1.41	0.87	0.001
Sequent	0.72	0.80	0.92	0.001	0.84	0.002
Hitachi	0.88	0.07	0.98	0.21	0.55	0.00
Apple	1.00	1.21	0.87	0.92	0.68	0.001
Data General	0.90	0.92	0.62	0.54	0.81	0.65

The DEA estimate of β_2^* is in units of the coefficient of the change in cost. The latter coefficient is close to one on the average.

show a very high elasticity of output from R&D spending. Sources of growth efficiency based on model shown in Equation (4.42) for the leading and nonleading firms on the efficiency frontier are as follows.

	Technical progress (%)		R&D efficiency (%)	
	1985–1989	1995–2000	1985–1989	1995–2000
Leading firms	25	30	28	30
Others	14	16	15	17

Now we consider a regression approach to specify the impact of R&D inputs on output. With net sales as proxy output (y) and x_1, x_2, x_3 as three inputs comprising R&D spending, net capital expenditure and all direct production inputs, we obtain

$$y = 70.8^* + 3.621^{**}\ x_1 + 0.291^{**}\ x_2 + 1.17^*\ x_3 \quad R^2 = 0.981$$

where * and ** denote significant t-values at 5 and 1 percent levels respectively. This uses a slightly reduced sample set. When the regressions are run separately for the DEA efficient and inefficient firms, the coefficient for R&D inputs is about 12 percent higher for the efficient firms, while the other coefficients are about the same. When each variable is taken in incremental form we obtain the result

$$\Delta y = -6.41 + 2.65^{**}\ x_1 + 1.05^{**}\Delta x_2 + 1.17^{**}\Delta x_3 \quad R^2 = 0.994$$

It is clear that the R&D variable has the highest marginal contribution to output (or sales), both in the level form and in the incremental form.

When we consider the DEA efficient firms only and several subperiods the regression results consistently show the dominant role of the R&D input in its contribution to sales as shown in Table 4.6.

The adjusted R^2 is very high, and the t-values for R&D expenditure are significant at 1 percent level. The elasticity of output with respect to R&D expenses estimated at the mean level comes out to 0.799 in 1985–1988 to 0.421 in 1985–2000.

There are two types of cost-efficiency models in DEA framework. One uses the cost frontier model to estimate the optimal level of efficient output y_j^* from Equation (4.30) obtained from the quadratic cost frontier. The second applies the dynamic cost frontier (Equation (4.44)) with average costs as the dependent variable. On applying the first model we compute AC_{min} from the optimal output y_j^* in Equation (4.30) defined by the quadratic cost frontier. The gap of the observed average cost from the minimum average cost may then measure the degree of underutilization of full capacity. Selected results for 1985, 1990, 1995 and 2000 are given in Table 4.7.

If we rank these selected companies by the gap $(AC - AC_{min})$ with the lowest gap allocated rank 1 and highest gap rank 12, the ranking in terms of cost-efficiency is as shown in Table 4.8.

Apple turns out to be the most efficient company at the beginning of the time period 1985–1990. At the end of 1990 the BCC model exhibited a decline from their leading position. Big companies like Toshiba did not exhibit a high degree of input efficiency and hence it was at the bottom

Table 4.6 Impact of R&D inputs for DEA efficient firms

	Intercept	x_1	x_2	x_3	$\bar{R}^2$
1985–1988	767.5	6.95**	1.38**	0.49	0.828
1993–1996	−146.6	2.54**	−0.09	1.35**	0.997
1997–2000	−239.9	4.00**	−0.15	1.19**	0.995
1985–2000	8.62	4.29**	0.11*	1.08**	0.996

Table 4.7 Degree of underutilization for selected companies

Company	1985		1990		1995		2000	
	AC	Gap	AC	Gap	AC	Gap	AC	Gap
Apple	0.65	0.0	0.59	0.0	0.85	0.19	0.81	0.12
Compaq	0.79	0.14	0.59	0.0	0.87	0.21	0.85	0.17
Datapoint	0.90	0.25	0.69	0.08	0.75	0.09	0.70	0.0
Dell	0.68	0.16	0.73	0.13	0.84	0.18	0.84	0.16
HP	0.90	0.25	0.84	0.23	0.82	0.16	0.83	0.16
Hitachi	0.99	0.34	0.96	0.36	1.01	0.35	1.11	0.43
IBM	0.91	0.26	0.94	0.33	0.87	0.21	0.88	0.20
Micron Electronics	0.70	0.20	0.80	0.20	0.86	0.21	0.89	0.21
Sequent	0.68	0.21	0.73	0.13	0.77	0.11	0.70	0.0
Silicon Graphics	0.69	0.04	0.63	0.0	0.64	0.0	0.90	0.22
Sun Microsystems	0.78	0.13	0.76	0.15	0.70	0.04	0.67	0.0
Toshiba	0.95	0.30	0.92	0.32	0.97	0.31	1.26	0.57

Table 4.8 Ranking of selected companies for selected years

Company	1985	1990	1995	2000
Apple	1	1	7	2
Compaq	4	8	9	4
Datapoint	9	3	3	1
Dell	5	5	6	3
HP	8	9	5	3
Hitachi	12	12	12	8
IBM	10	11	9	5
Micron	6	7	8	6
Sequent	7	4	4	1
Silicon Graphics	32	2	1	7
Sun	3	6	2	1
Toshiba	11	10	11	9

of the efficiency ranking. Sun Microsystems tends to outperform the other companies on average regarding cost-efficiency. Recently it has kept up its dominant position in the ladder. Thus our cost-based DEA model identifies the MES for each efficient firm's cost frontier and hence the gap analysis is useful in identifying the degree of underutilization of full capacity.

For the second application the dynamic average cost frontier may be specified as

$$\frac{\Delta c_j(t)}{c_j(t)} = \beta_0^* + \beta_1^* \left(\frac{\Delta y(t)}{y(t)} \right) - \beta_2^* \left(\frac{\Delta x_1(t)}{x_1(t)} \right)$$

where the values of β_2 measure the cost-reducing impact of growth in R&D spending. For selected companies the results are as follows:

Company	**1985–1988**		**1988–1991**		**1997–2000**	
	β_2^*	θ^*	β_2^*	θ^*	β_2^*	θ^*
Apple	1.21	1.00	1.26	0.90	0.001	0.87
Compaq	0.03	0.97	1.50	1.00	0.04	0.95
Hitachi	0.07	0.88	0.04	1.00	0.002	0.55
IBM	2.82	1.00	1.61	1.00	0.71	1.00
Toshiba	1.56	0.93	0.04	0.84	0.05	0.79

Clearly the R&D spending contributes significantly to the growth efficiency of DEA efficient firms. Thus the learning-by-doing effect is so important for modern industries.

4.5.3 Efficiency in pharmaceutical industry

We have used the same Standard and Poor's Compustat Database to analyze the efficiency structure in the pharmaceutical industry (PI). The pharmaceutical companies have grown immensely over the recent past, as breakthroughs in recent medical research have led to the development of new medicines and procedures. An overview of growth of demand and the direct production costs, usually termed "cost of goods sold", was obtained for five selected companies over the period 1982–2000 by the first-order autoregressive equation

$$y_t = a + by_{t-1} \tag{4.45}$$

where y_t denotes net sales or cost of goods sold. The five companies are Abbott Lab, Bausch and Lomb (B&L), Merck, Pfizer and Pharmacia Corporation. Table 4.9 reports the estimates.

Here * and ** denote significance of t-values at 5 percent and 1 percent respectively. Note that net sales growth rate measured by $\hat{\beta} = \hat{b} - 1$ has been the highest for Pfizer (0.553), followed by Pharmacia (0.406), Merck (0.220) and Abbott (0.042). These growth rates are all statistically

Table 4.9 Autoregressive estimates of net sales and cost of goods sold for the pharmaceutical industry (1981–2000)

	Net sales				Cost of goods sold			
	$\hat{a}$	$\hat{b}$	$\bar{R}^2$	DW	$\hat{a}$	$\hat{b}$	$\bar{R}^2$	DW
Abbott	313.71**	1.042**	0.997	1.05	27.76	1.070	0.996	2.47
B & L	175.21	0.915**	0.896	2.48	92.11	0.866**	0.808	2.48
Merck	−448.2	1.220**	0.994	2.67	−65.69	1.269**	0.985	1.90
Pfizer	−2649.5	1.553**	0.260	1.136	28.91	0.909	0.044	0.982
Pharmacia	−2579.6	1.406**	0.260	0.746	1954.3*	0.527**	0.261	1.319

significant at 1 percent level. If we measure the growth rate of net sales per net production cost by $g = \Delta y/y - \Delta c/c$, then the highest growth rate is exhibited by Pharmacia ($g = 117\%$) and Pfizer ($g = 64\%$). Thus the profit growth for these two companies is found to be very high.

We now consider a larger set of companies for assessing the impact of R&D investments in research and innovations over 19 years (1981–2000). A set of 17 companies out of a larger set of 45 is selected from the Compustat database available from Standard and Poor. This selection is based on considerations of continuous availability of data on R&D expense and its share of total costs. The selected companies comprise such well-known firms as Merck, Eli Lily, Pfizer, Bausch and Lomb, Johnson and Johnson, Glaxosmithkline, Schering-Plough and Genentech. The share of R&D in total costs is quite important for these companies.

The distribution of net sales over the period 1981–2000 for these 17 companies is as follows:

	1981	1990	2000
Mean	2061.07	4263.24	4263.24
Standard deviation	2125.07	383.69	518.22
Skewness	0.8317	0.2973	0.7458

Clearly the data are more homogenous in years 1990 and 2000 compared to the year 1981.

Four types of estimates are calculated for the selected companies in the pharmaceutical industry. Table 4.10, provides the estimates of cost-efficiency along the total cost frontier. The model here is of the form

$$\text{Min } \theta$$

$$\text{subject to } \sum_{j=1}^{n} C_j\lambda_j \le \theta C_h; \sum_{j=1}^{n} x_j\lambda_j \le x_h$$

$$\Sigma y_j\lambda_j \ge y_h; \Sigma\lambda_j = 1, \lambda_j \ge 0 \tag{4.46}$$

$$j = 1, 2, \ldots, n$$

where firm h is the reference firm with output y_h and costs C_h and x_h. Here x_h is R&D costs and C_h is total costs excluding R and R costs. Total costs comprise cost of goods sold, net plant and machinery expenditure and all marketing costs excluding R&D expenses denoted by x_h. A growth efficiency form of this model has been analyzed before. The optimal values of the LP model denoted by * are such that

$$\theta^* = 1 \text{ with } \Sigma C_j\lambda_j^* = C_h; \quad \Sigma\ \lambda_j^* = x_h$$

Then the firm h is efficient, that is it lies on the cost-efficiency frontier; also the R&D inputs are optimally used. If, however, $\theta^* < 1$, then $\sum C_j\lambda_j^* < C_h$, indicating that optimal costs $C_h^* = \sum C_j\lambda_j^*$ are lower than the observed costs C_h. Hence the firm is not on the cost-efficiency frontier.

A second type of estimate uses the growth efficiency model to characterize the efficient and inefficient firms and then applies the regression model in order to estimate the impact of the growth of R&D inputs. This is compared with the level effect, when we regress total cost on R&D and other variables. A third type of estimate calculates the impact of e and other component inputs on total sales revenue for firms which are on the cost-efficiency frontier.

Finally, we estimate the market share models where for cost-efficient firms it is tested if their market share has increased when the R&D inputs helped reduce their average costs.

Table 4.10 reports the optimal values θ^* of the LP model (Equation (4.46)) for each firm for three selected years 1981, 1990 and 2000. If instead of TC, we use AC defined by the ratio of total costs to net sales, the estimates of θ^* change but not very significantly. Table 4.11 presents a summary of firms which are efficient in terms of TC, AC and RD level.

Table 4.10 Efficiency coefficients (θ^*) for the total cost and the average cost frontier

	1981		1990		2000	
	TC	AC	TC	AC	TC	AC
Abbott Lab	0.829	0.831	0.871	0.885	0.807	0.832
Alza Corp	0.312	0.324	0.452	0.453	0.800	0.802
American Home Products	1.000	1.000	1.000	1.000	0.771	0.809
Bausch & Lomb	0.877	1.000	0.768	1.000	0.737	0.739
Bristol Myers	0.832	0.861	0.971	1.000	0.939	0.982
Forest Lab	0.878	1.000	0.661	0.662	0.531	0.532
Genentech	0.264	0.273	0.549	0.559	0.545	0.556
Glaxosmith	0.493	0.514	0.787	0.818	0.847	0.964
IGI Inc	1.000	0.024	1.000	0.709	1.000	1.000
Johnson & Johnson	0.958	1.000	1.000	1.000	0.938	1.000
Eli Lily	0.772	0.886	0.781	0.811	0.840	0.903
Merck	0.710	0.848	0.983	1.000	1.000	1.000
MGI Pharma	0.548	0.196	1.000	0.199	0.680	0.442
Natures Sunshine	0.432	1.000	1.000	1.000	1.000	1.000
Pfizer Inc	0.764	0.822	0.838	0.822	0.841	1.000
Schering-Plough	0.703	0.709	0.796	0.808	0.837	0.872

Table 4.11 Number of efficient firms with efficient TC, AC and R&D

	TC	AC	R&D level
1981	3 (18%)	6 (35%)	6 (35%)
1990	5 (29%)	6 (35%)	6 (35%)
2000	3 (18%)	5 (29%)	5 (29%)

Tables 4.12 and 4.13 report the estimates of the cost frontier in two forms: the level form and the growth form respectively, where the inputs are separately used as an explanatory variable. Here the growth form exhibits much better results over the level form.

Finally, Table 4.14 reports the estimates of the market share models, where each firm is analyzed over the whole period. The market share model predicts that the efficient firms would increase their market shares when the industry average cost rises due to the failure of inefficient firms to reduce their long-run average cost. For the whole industry over the period 1981–2000 this relationship is tested by the following regressions,

Table 4.12 Cost frontier estimates of selected firms over the whole period 1981–2000, $TC_j = a + by_j + C\hat{R}_j + d\theta_j$

Firm	*a*	*b*	*c*	*d*	R^2	*F* statistics
ABT	1355**	1.301**	−295.3*	−0.046[a]	0.999	3933.2
AHP	15410	0.645	−5.8E07	−13068.7	0.999	4402.3
BOL	−250.3	1.608**	N	−1375.4*	0.996	780.22
GSK	1915.1**	1.335**	−10929.1	−1505.7*	0.993	378.4
IG	−37.17	3.112**	245.1	−1.510[a]	0.987	221.1
PF	1672.7**	1.350**	−46652.7**	−0.064[a]	0.999	4196.6
PHA	−16807.7*	2.224**		−1.517*[a]	0.985	187.2

* and ** denote significance of *t* values at 5 and 1 percent respectively; the superscript "a" denotes the cross-product of efficiency and output levels as the repressor, since the output term was highly dominant. TC_j, y_j and θ_j are total cost, output and efficiency scores. $\hat{R}_j$ is a proxy for R&D combined with output; *N* denotes a high value which is not significant at even 20 percent level of *t* test; for other firms not included here multicollinearity yields singularity of estimates, hence these are not reported.

Table 4.13 Sources of growth of total costs for the industry as a whole, $GTC_j = a + bGRD_j + Gy_j$

	1982	1991	2000
a	−0.124	0.097**(*D*)	−0.165**
b	0.914**	0.389**(*D*)	0.815**
c	6.15E−06	−6.2E − 06*	9.11E−06**
R^2	0.871	0.653	0.913
F	47.21	13.176	73.424

GTC, GRD and Gy denote the proportional growth rates of total costs, R&D and total output respectively; * and ** denote significant *t* values at 5 and 1 percent respectively; *D* denotes a dummy variable with one for the efficient units and zero for others. It indicates that these coefficients are significantly different for the efficient firms compared to the inefficient ones.

where $\bar{c}$ is the industry average cost function including both efficient and inefficient firms.

$$\bar{c} = \underset{(t=10.02)}{3953.2}^{**} + \underset{(8.323)}{5.79}^{**}\ \sigma^2; \quad R^2 = 0.912,\ F = 186.98$$

$$\Delta\bar{c} = \underset{(t=2.31)}{0.145}^{**} + \underset{(3.45)}{0.005}^{**}\ \sigma^2 - \underset{(1.91)}{0.014}^{*}\ \bar{c}; \quad R^2 = 0.721,\ F = 291.01$$

Note that the coefficient of variance σ^2 in the first equation is in units of E − 05. Thus higher variance tends to increase average industry costs

Table 4.14 Estimates of market share models for selected firms in the pharmaceutical industry (1981–2000), $\Delta s = b_0 + b_1(\bar{c} - c(u))$

Firm	b_0	b_1	R^2	F
ABT	1.07**	2.267**	0.385	5.002
AHP	1.295**	4.299	0.155	1.464
BOL	1.095**	0.691[a]	0.296	3.156
BM	1.022**	1.577*	0.224	2.312
GSK	1.082**	0.727	0.027	0.221
PF	1.209**	12.271**[a]	0.492	7.750
PHA	1.121**	3.947*	0.198	1.980

* and ** denote significance of t values at 5 and 1 percent respectively; the superscript "a" denotes that the quadratic term $(\bar{c} - c(u))^2$ has a significant positive coefficient.

and allows efficient firms to increase their market share. Note that the impact of variance is highly significant statistically. Clearly the churning effect is found to be important for this industry. Thus higher variance tends to increase average costs and allows efficient firms to increase their market share. Note that the impact of variance is highly significant statistically.

Several points emerge from the estimated results in Tables 4.10–4.14. First of all, the number of firms on the cost-efficiency frontier is about one-third and these firms are invariably efficient in using their R&D inputs. Secondly, both the efficiency score and the composite R&D inputs help the firms improve their cost-efficiency and these results are statistically significant. Growth of R&D inputs is as important as output growth in contributing to the increase of costs over time. This implies that the R&D inputs play a very dominant role in the growth of the pharmaceutical industry. It also increases profit through higher demand. Thirdly, the market share model shows very clearly that the more efficient firms with $\bar{c} > c(u)$ increase their market share over time and the two sources of this share gain are the decrease in average cost through R&D and other forms of innovation and the increase in industry-wide average cost due to the failure of less efficient firms to reduce their long-run average costs. Clearly when the cost heterogeneity measured by cost variance σ^2 rises, it tends to increase the industry average cost ($\bar{c}$) over time. This creates a long-run force for increased entry and/or increased market share. We now consider this long-run process of industry evolution over time due to increasing core competence of the DEA efficient firms.

4.5.4 Core competence and industry evolution

What makes a firm grow? What causes an industry to evolve and progress? From a broad standpoint two types of answers have been offered. One is managerial, the other economic. The managerial perspective is based on organization theory, which focuses on the cost competence as the primary source of growth. The economic perspective emphasizes productivity and efficiency as the basic source of growth. Economic efficiency of both physical and human capital including innovations through R&D have been stressed by the modern theory of endogenous growth.

Core competence rather than market power has been identified by Prahalad and Hamel (1994) as the basic cornerstone of success in the modern hypercompetitive world of today. Core competence has been defined as the collective learning of the organization, especially learning how to coordinate diverse production skills and integrate multiple streams of technologies. Four basic elements of core competence are as follows: learn from own and outside research, coordinate, integrate so as to reduce unit costs and innovate so as to gain market share through price and cost reductions.

A company's own R&D expenditures help reduce its long-run unit costs and also yield spillover externalities. These spillovers yield IRS as discussed before. Now we consider a dynamic model of industry evolution, where R&D investments tend to reduce unit costs and hence profitability. This profitability induces new entry and also increased market share by the incumbent firms who succeed in following the cost-efficiency frontier.

Denoting price and output by p and y respectively, the dynamic model may be specified as

$$\dot{y} = a(p - c(u)), \quad a > 0 \tag{4.47}$$

where dot denotes time derivative and $c(u)$ is the average cost depending on innovation u in the form of R&D expenditure. When total profit $\pi = (p - c(u)),\ y$ is positive, it induces entry in the form of increased output over time. Entry can also be represented by $\dot{n}$, where n denotes the number of firms, but we use $\dot{y}$ since n is discrete. We assume that each incumbent firm chooses the time path $u(t)$ of R&D that maximizes the present value $v_0 = \int_0^\infty \exp(-rt)\pi(u)dt$ of future profits as the known discount rate r. The current value of profits at time t is

$$v(t) = \int_t^\infty \exp(-r(\tau - t))\pi(u)d\tau$$

On differentiating $v(t)$ one obtains

$$\dot{v} = rv - \pi(u) \tag{4.48}$$

This represents capital market efficiency or the absence of arbitrage.

The dynamic model defined by Equations (4.47) and (4.48) is a model of industry evolution. When excess profits is zero, one obtains the equilibrium $p = c(u^*) = c(y^*, u^*)$. Again if the cost of entry equals the net present value of entry $v(t)$, then $\pi(u^*) = rz$, z being the cost of entry with $z = v$. The dynamics of the evolution model can be discussed in terms of a linearized version of Equations (4.47) and (4.48) and the associated characteristic roots. Sengupta (2004) has analyzed the stability aspects of this dynamic model elsewhere.

The profitability equation (4.17) may also be written in terms of the market share s of the incumbent firm as

$$\dot{s} = b(\bar{c} - c(u)) \tag{4.49}$$

when price is assumed to be proportional to the industry average cost function $\bar{c}$, where $\bar{c}$ is the average of both best practice firms and others. Mazzacato (2000) and Sengupta (2004) have applied this type of dynamics of market evolution in several industries. Whenever $\bar{c} > c(u)$ the incumbent firm increases its market share. Also by the optimal allocation of R&D innovations u, the incumbent firms may succeed in reducing unit costs $c(u)$ in the long run. This also increases their market shares. Following the Fisherian model of growth of fitness in natural evolution of species, Mazzacato (2000) has shown that the rate of change in industry average cost function may be viewed as proportional to the variance of individual costs $c_i(u_i)$ so that

$$\frac{d\bar{c}}{dt} = \dot{\bar{c}} = \alpha\sigma^2(t)$$

More generally it may be written as

$$\dot{\bar{c}} = \alpha_1\sigma^2(t) - \alpha_2\bar{c} \tag{4.50}$$

where α_1, α_2 are nonnegative coefficients and α_0 is the intercept term. Thus if cost variances rise, the industry average cost rises, implying a fall in overall efficiency. On the other hand, if $\bar{c}$ rises, it tends to reduce the growth rate of $\bar{c}$ over time due to more exits. This impact of heterogeneity in costs has sometimes been called the "churning effect" by Lansbury and Mayes (1996), who analyzed the entry–exit dynamics of several industries in the United Kingdom.

We have to note that the concept of core competence depending on overall cost-efficiency of firms is closely related to the efficiency of R&D investments and their impact on learning by doing. One way to capture this impact is to reformulate the DEA model as a profit-maximizing model for choosing optimal output y, average cost (c) and R&D spending R when the output price p is given; for example

$$\max\ \pi = py - cy - R$$

$$\text{subject to}\quad \sum_{j=1}^{n} c_j\lambda_j \le c,\quad \sum_j R_j\lambda_j \le R,\quad \sum_j y_j\lambda_j \ge y$$

$$\Sigma\lambda_j = 1,\quad \lambda_j \ge 0;\quad j = (1, 2, \ldots, n)$$

On using the Lagrangean function

$$L = py - cy - R + \beta(c - \sum_j c_j\lambda_j) + b(R - \Sigma R_j\lambda_j) + \alpha(\sum_j c_j\lambda_j - y) + \beta_0(\Sigma\lambda_j - 1)$$

where β_0 is free in sign, we can compute for the efficient firm j with positive levels of c, r and y:

$$p = c^* + \alpha^*,\quad b^* = 1,\quad \beta^* = y^* \tag{4.51}$$

This yields the total cost frontier as

$$\beta * c_j = C_j^* = \beta_0^* + \alpha^* y_j - b^* R_j \tag{4.52}$$

This assumes that R does not affect average cost c. However, if it does, then the cost frontier (Equation (4.52)) reduces to

$$\frac{R^*}{C^*} = \frac{\varepsilon_R}{(1 - b^*)} \tag{4.53}$$

where ε_R denotes the R&D elasticity of average cost, that is $\varepsilon_R = -(\partial c/\partial R)/(c/R)$. Thus the optimal R&D share of total costs is proportional to the R&D elasticity of demand since b^* is usually less than one, and higher R&D elasticity representing higher cost-reducing effect tends to increase optimal R&D spending.

When we add the quadratic constraints on output and R&D spending on

$$\sum_{j=1}^{n} y_j^2\lambda_j \ge y^2,\quad \sum_j R_j^2\lambda_j = R^2 \tag{4.54}$$

we obtain the dynamic cost frontier in terms of optimal average cost c_1^* for the j-th efficient firm:

$$c_j^* = \left(\frac{1}{\beta^*}\right)(\beta_0^* + \alpha_1^* y_j + \alpha_2 y_j^2 - b_1^* R_j + b_2^* R_j^2)$$

where β_0^* and b_2^* are free in sign. In this case the MES can be determined by optimally choosing the output and R&D levels whenever it is meaningful.

Note that the cost frontier model defined by Equations (4.51) and (4.52) can be easily applied to the whole industry of efficient firms. Let all firms be efficient and total industry output be denoted by $Q^* = \sum_{j=1}^{n} y_j$. Then one obtains in equilibrium

$$\pi^* = py^* - c^* y - R^* = 0$$

Adding over n^* efficient firms one obtains

$$\frac{n^* R^*}{pQ^*} = 1 - \frac{c^* Q^*}{pQ^*} = \frac{1}{n^* \varepsilon_\text{p}} \tag{4.55}$$

Since $p = c^* = p/(n^* \varepsilon_\text{p})$, where ε_p is price elasticity of demand or the reciprocal of Lerner index of degree of monopoly. Since a rise in n^* yields a lower industry-level R&D to sales ratio, it is more likely that industries with naturally more competitive structure will do less R&D effort, all else being equal. Similarly higher price elasticity, which is a competitive feature, will tend to lower R&D to sales ratio. In case the industry does not have all firms equally efficient, that is some firms do not adopt the optimal values c^*, y^*, R^*, the net outcome would be slightly different as in monopolistic competition with product differentiation. Thus n_1^* firms ($n_1^* < n$) will follow the efficiency rules (Equation (4.55)) but others may not.

4.5.5 Industry evolution under innovations

To analyze industry evolution we adopt a two-stage formulation. Average cost is used as a measure of size and the first stage applies the DEA model to identify a subset of cost-efficient firms. In the second stage we analyze the growth of efficient firms which increases their market share by cost-reducing strategies based on R&D investments. The von Neumann model defined dynamic efficiency in terms of optimal rates of expansion. We apply here a dynamic model of industry evolution.

We consider an industry consisting of homogenous firms competing in R&D investments. The expected net cost of a representative incumbent firm is assumed to be a function of the number of firms n in the industry and of the R&D expenditure u, that is $c(n, u)$. The number of firms n is assumed to be known to all incumbent firms at the beginning of time period $(t, \ t+\mathrm{d}t)$. The R&D parameter u is the effort made by the firm in product innovation at time t.

We make two specific assumptions for obtaining specific results.

Assumption 1 For each positive n the long-run cost function $c(n, u)$ has an interior minimum in u. This yields $c(n) = \arg \underset{u}{\mathrm{Min}} \ c(n, u)$. This assumption reflects diminishing returns to R&D at the firm level.

Assumption 2 New entry ($\dot{n} = \mathrm{d}n/\mathrm{d}t$) occurs whenever profits $\pi(n) = p - c(n)$ are positive. This assumption reflects the incentives behind new entry, that is exit occurs when profits are negative. For simplicity we assume that the entry and exit processes are linear functions of expected profits $\pi(n)$.

Since n is discrete, we replace it by output y, so that the entry–exit dynamics can be modeled as

$$\dot{y} = a(p - c(y)), a > 0 \tag{4.56}$$

In a static model a free entry condition would simply imply that the output in equilibrium is such that $p = c(y)$. This yields the equilibrium number of firms in the industry.

Each incumbent firm in this competitive framework is a price taker and therefore minimizes the initial present value of long-run costs $C_0 = \int_0^\infty \exp(-rt)c(y, u)\mathrm{d}t$ in order to stay in the industry indefinitely. At any time $t > 0$ the incumbent firm's average cost is given by

$$C(t) = \int_t^\infty \exp[-r(\tau)]c(y, u) \ \mathrm{d}t$$

Differentiating this cost function one obtains

$$\dot{C} = rC - c(y, u) \tag{4.57}$$

Equations 4.47 and 4.48 specify the dynamic model of industry evolution under competitive conditions. This is called the cost model. Here the dynamic decision problem for the incumbent firm is to choose the time path of the R&D investments $u = u(t)$ which minimize the initial cost $C_0 = C(0)$ subject to the state equations (4.56) and (4.57). The dynamic problem for the incumbent firm is to select the time path of the R&D expenditure $u(t)$ such that it minimizes the initial discounted cost $C_0 = C(0)$ subject to the state Equations (4.56) and (4.57). The current value Hamiltonian is

$$H = c(y,u) + s_1(a(p - c(y,u))) + s_2(rC - c(y,u)) \tag{4.58}$$

where s_1 and s_2 are the respective costate variables associated with (4.56) and (4.57). Clearly the optimal control function minimizes the expression $(1 - s_2)\, c(y,u)$ and by Assumption 2 the optimal control is $c(y)$. The costate variable s_1 is the shadow price of the new firm and it is positive for all t if its initial value $s_1(0)$ is chosen in a proper way, since

$$\dot{s}_1 = rs_1 - (1 - s_2)c'(y), c'(y) = \frac{\partial c}{\partial y}$$

Note that the optimal control does not depend on the costate variables, so that we may ignore the costate variables and focus on the dynamics defined by the state variables y and C. This is the system

$$\dot{y} = a(p - c(y)) \tag{4.59}$$

$$\dot{C} = rC - c(y) \tag{4.60}$$

When $p = c(y)$ there is no new entry into (or exits from) the industry in the long run. The steady state output y^* and the associated number n^* of firms in the industry are given by the solution of the equation

$$p = c(y) = rC \tag{4.61}$$

Clearly the shape of the average cost function $c(y)$ determines the nature of evolution of firms in the industry. The steady state cost C^* is given by

$$C^* = \frac{c(y)}{r}$$

On linearizing the dynamic system (Equations (4.59) and (4.60)) the characteristic equation may be written as

$$\lambda^2 + (ac' - r)\lambda - rac' = 0$$

with eigenvalues

$$\lambda = \left(\frac{1}{2}\right)\left[(r - ac') \pm \left((r - ac')^2 + 4rac'\right)^{1/2}\right] \tag{4.62}$$

Several cases may be analyzed. In case $c' = \partial c/\partial y$ is positive and the eigenvalues are real, then one eigenvalue is positive and one negative. This yields the stationary state to be a saddle point. When c' is negative and a is sufficiently small, both characteristic roots are positive and the stationary equilibrium is an unstable node. Finally, if c' is negative and $(r - ac')^2 < 4rac'$, then the roots are complex-valued with positive real parts. In this case the stationary state is an unstable focus, where cyclical evolutions persist. Thus the equilibrium path of industry evolution is a trajectory of the dynamic system (Equations (4.59) and (4.60)) around any of the three long-run equilibria. Given the initial number of firms n_0 with associated output y_0, the equilibrium path is determined by the initial cost C_0. The cyclical dynamics imply a multiplicity of equilibrium paths. For a given initial number of firms n_0 with output y_0, there may be many initial values C_0 specifying the movement toward one of the steady states.

In case of saddle point stationary state there is a stable manifold along which motion is purely towards (y^*, C^*) and an unstable manifold along which motion is exclusively away from (y^*, C^*), and these are given by the eigenvectors of the coefficient matrix of the system (Equations (4.59) and (4.60)) in a linearized form corresponding to the stable and unstable roots respectively.

The growth model in Equations (4.59) and (4.60) can be easily modified to an investment model which assumes that the unit costs may be reduced by oligopolistic firms choosing higher scales of plant. This model may be rewritten in terms of average industry cost $\bar{c}$ and its deviation from the innovating firms' cost $c(k)$, that is

$$\dot{y} = a(\bar{c} - c(k)) \tag{4.63}$$

$$\dot{k} = I - \delta k$$

Thus firms grow in size if $c(k)$ is less than $\bar{c}$. By improving cost-efficiency a firm can grow faster. This type of model is closely related to the dynamic evolution model developed by Mazzacato (2000), who argued that the cost reduction process, also called the dynamic increasing returns, may occur at diverse rates for different firms thus increasing the comparative advantages of the successful firms and decreasing the same for the laggards. Two interesting implications follow from this type of evolution model. First, the major source of growth here is the productivity gain or efficiency. Any means which help improve efficiency would improve the growth in size measured by output. This has been empirically supported by several studies; for example, Lansbury and Mayes (1996) have found for industrial data in the United Kingdom that the entry and exit processes are mainly explained by the rise and fall of productive efficiency respectively. Secondly, this provides the basis of the modern evolutionary theory of competition, which is closely related to the competitive fitness model of growth of biological species.

Sengupta (2004) has analyzed in some detail this investment model to explain the market dynamics of the evolution of industry and the impact of new technology with R&D and the knowledge capital.

Thus we may conclude that production efficiency and R&D investments are the major determinants of industry evolution today in the high-tech fields such as computers, electronics and pharmaceuticals. A dynamic view of DEA efficiency models must contain two components. The first is the overall cost-efficiency as analyzed by the DEA models of technical and allocative efficiency and the second is the optimal rate of expansion of firms which are on the cost frontier in the first stage. The two components are seen here as mutually complementary.

References

Aghion, P. and P. Howitt (1992) A model of growth through creative destruction, *Econometrica* 60, 323–351.

Aghion, P. and P. Howitt (1998) *Endogenous Growth Theory* (Cambridge: MIT Press).

Brander, J. and B. Spence (1983) Strategic commitment with R&D: The symmetric case, *Bell Journal of Economics* 14, 225–235.

Brown, M. and S. Chiang (2003) *Coalitions in Oligopolies* (Amsterdam: Elsevier).

Buckley, P. (2000) Foreign market entry: A formal extension of internalization theory, in M. Casson (ed.) *Economics of International Business* (Cheltenham: Edward Elgar).

Caves, R. and M. Porter (1997) From entry barriers to mobility barriers, *Quarterly Journal of Economics* 9, 241–267.

Geroski, P. (2003) *The Evolution of New Markets* (Oxford: Oxford University Press).

Geroski, P. and R. Masson (1987) Dynamic market models in industrial organization, *International Journal of Industrial Organization* 5, 1–14.

Ishii, Y. (2000) International Cournot duopoly with R&D subsidies under demand uncertainty, *Journal of Economics* 72, 203–222.

Kamien, M. and N. Schwartz (1982) *Market Structure and Innovation* (Cambridge: Cambridge University Press).

Lansbury, M. and D. Mayes (1996) *Sources of Productivity Growth* (Cambridge, UK: Cambridge University Press).

Lukach, R., J. Plasmans and P. Kort (2005) Innovation strategies in a competitive dynamic setting. Working Paper No. 1395, Center for Economic Studies, Munich, Germany.

Masson, R. and J. Shaanan (1982) Stochastic dynamic limit pricing: An empirical test, *Review of Economics and Statistics* 64, 413–423.

Matsuyama, K. and M. Itoh (1985) Protection policy in a dynamic oligopoly market. Working paper, University of Tokyo, Department of Economics, Tokyo.

Mazzacato, M. (2000) *Firm Size, Innovation and Market Structure* (Cheltenham: Edward Elgar).

Ponssard, J. (1979) The strategic role of information on the demand function in an oligopolistic market, *Management Science* 25, 243–250.

Prahalad, C.K. and G. Hamel (1994) *Competing for the Future* (Cambridge: Harvard University Press).

Sengupta, J.K. (2003) *New Efficiency Theory: With Applications of Data Envelopment Analysis* (New York: Springer).

Sengupta, J.K. (2004) *Competition and Growth: Innovations and Selection in Industry Evolution* (New York: Palgrave Macmillan).

Tirole, J. (1988) *The Theory of Industrial Organization* (Cambridge: MIT Press).

Veloce, W. and A. Zellner (1985) Entry and empirical demand and supply analysis, *Journal of Econometrics* 30, 459–471.

5
Industry Evolution under Entry Barriers

5.1 Introduction

The evolution of industry depends on the selection mechanism, that is the process of entry and exit of firms and the various factors influencing the entry–exit decisions. Two major factors are most important in the selection processes. One is the evolutionary perspective which emphasizes the firm's ability through strong increasing returns to scale to alter the market structure significantly. Following the Schumpeterian theory of innovations the main source of such increasing returns is the cumulative aspect of such innovations. Secondly, firms differ significantly in their commitment and ability to innovate. Thus innovations in products and processes are largely endogenous to the firm through R&D investment and learning by doing. Thus vigorous innovation has been found to generate more competitive market structures, while innovations requiring large investment generally involve more concentrated market structures. Thus Klepper (1996) and others have shown that innovations could lead to increased concentration if successful innovators rise to market dominance by causing the MES of production to grow more rapidly than demand.

One of the major determinants of the market entry process is the existence of potential and actual barriers. The pioneering work by Bain (1956) argued that firms can earn profits above the competitive level in the medium and the long term if they are protected by entry barriers. He identified four types of entry barriers, for example economies of scale, product differentiation advantages, absolute cost advantages and large capital stock requirements. To this list one may add two more: innovations efficiency in the timing and know-how of new investments and creating strongholds for excluding competitors from their market segments. These latter two have been emphasized by D'Aveni's (1994) theory of hypercompetition.

The evolution of firms in an industry follows a competitive adjustment process when there exist no entry and exit barriers. Dreze and Sheshinski (1984) have considered such dynamic adjustment processes in the Walrasian tradition, which assume that new plants of a given type are built (scrapped) whenever their expected profits are positive (negative). The presence of entry barriers makes the market framework oligopolistic, where the Cournot–Nash equilibria provide a reasonable solution. Suzumura (1995) has analyzed this framework and developed an entry dynamics in the form of an increase in the number of firms proportional to the excess profits above the Cournot–Nash equilibrium. He has shown that the prevalence of entry barriers, which protect the incumbent firms from threats by potential competitors is not necessarily decreasing social welfare, where the latter is measured by the total of CS and PS in a competitive system. Gaskins (1971) and later Judd and Petersen (1986) have developed a two-player differential game model with two groups of firms: the new entrants and the incumbents. In this type of model positive entry in the form of an increase in output of the entrants occurs as soon as the market price (of the incumbent firm) exceeds the limit price, where the latter may be viewed as the price at which entry is zero. Clearly the limit price is very close to the competitive equilibrium price which equals minimum average cost. Thus the incumbent here is either a dominant firm or a group of joint profit-maximizing oligopolists facing expansion by the entrants who comprise the firms in the competitive fringe. Note that the limit pricing model does not consider entry in the form of an increase in the number of firms ($\dot{n} = \mathrm{d}n/\mathrm{d}t$); entry is viewed here as an expansion of output by the competitive fringe.

This chapter studies the stability of the entry and exit process and the industry evolutions in both competitive and noncompetitive markets. It first surveys the competitive adjustment mechanism in the traditional Walrasian framework and presents three sources of instability, for example adjustment lag, increasing returns to scale and the entry–exit dynamics in the competitive process. Then this chapter changes the scenario and discusses the noncompetitive market structures, where Cournot–Nash type equilibria prevail. It is shown that the entry barriers, potential or actual, generated by large capital investment and/or the R&D investment cost may result in unstable manifolds in the adjustment process exhibiting cyclical fluctuations. We also address the case in which the incumbent firm can precommit to the level of output and thereby show that the adjustment process which contains both entry deterrence and accommodation can acquire stability characteristics.

Our object here is twofold. We discuss the adjument process underlying the entry dynamics when there exist potential or actual entry barriers and their stability characteristics over time, when entry costs compete with the expected profitability of entering the market, which is oligopolistic.

5.2 Competitive adjustments

Competitive adjustment mechanisms in the Walrasian framework assume free entry with no entry barriers and also flexible prices and quantities. Long-run entry decisions are motivated by expected profitability, when all costs are variable, that is, only in the short run some costs are fixed but entry decisions assume all costs as variable. Non-Walrasian adjustments incorporate the presence of entry barriers in some form, sticky or fixed prices due to delayed quantity adjustment and market disequilibrium which may involve rationing or inventory costs and also delayed response. In this section we discuss in brief the entry dynamics under competitive adjustment.

Three forms of entry dynamics under competitive adjustment have been used in the current literature, each with its own stability characteristics. One assumes that prices rise in response to excess demand and fall in response to excess supply, and firms' output adjust according to profitability. This is the classical Walrasian adjustment process analyzed in some detail by Arrow and Hurwicz (1962), Morishima (1964), Dreze and Sheshinski (1984), Heal (1986), Novshek and Sonnenschein (1986) and many others. The interpretation of this Walrasian dual adjustment process when in disequilibrium is as follows: if the planned demand $d(p)$ of a commodity is larger (smaller) than the planned supply y, then the tentative price (p) increases (decreases), that is

$$p = b(d(p) - y) \tag{5.1}$$

$$\dot{y} = a(p - c), \quad c = c(y), \quad a > 0 \tag{5.2}$$

The equilibrium of the system (5.1) and (5.2) is defined by the values (p^*, y^*) at which $\dot{p} = 0 = \dot{y}$.

A second form of the entry dynamics under competitive adjustment considers only one of the two adjustments, either price in Equation (5.1) or quantity in Equation (5.2). Thus Dreze and Sheshinski (1984) consider entry in terms of the rate of change of number of firms ($\dot{n} = \mathrm{d}n/\mathrm{d}t$) depending on expected profitability. They assume that each firm produces a homogeneous output y_i with a total cost $C_i(y_i)$, such that each firm is assumed to belong to one of K possible types of the cost structure. Then we have n_i as the number of firms of cost type $i = 1, 2, \ldots, K$. Each cost function is assumed to be monotone increasing, strictly convex and twice differentiable with a standard U-shaped average cost curve. The industry equilibrium is then obtained by solving the model

$$\begin{aligned} &\underset{n_i y_i}{\mathrm{Min}}\ C = \sum_{i=1}^{K} n_i C_i(y_i) \\ &\text{subject to} \sum_{i=1}^{K} n_i y_i \geq d, \quad d > 0 \end{aligned} \tag{5.3}$$

where d is total market demand assumed to be exogenous. On using p as the optimal shadow price of the demand–supply inequality, one easily derives the optimal solutions (n_i^*, y_i^*) by Kuhn–Tucker theorem, where it holds for positive n_i^* and y_i^* that

$$p = C_i' = \mathrm{MC}_i = \frac{C_i}{y_i^*} = \mathrm{AC}_{\min}$$

Sengupta (2003) has discussed the short- and long-run implications of this equilibrium condition, when the capital inputs are fixed (short run) or variable (long run). Given the optimal solutions $n^* = (n_i^*)$, $y^* = (y_i^*)$ at positive levels the dynamic process of entry and exit of firms is modeled by Dreze and Sheshinski as follows:

$$n_i^* = g_i(\pi_i) \tag{5.4}$$

where $\pi_i = py_i - C_i(y_i)$ is total profits, $g_i(\cdot)$ shows continuous and strictly increasing functions with $(0) = 0$ for all $i = 1, 2, \ldots, K$ and $d = \Sigma n_i y_i$. On linearizing Equation (5.4) the entry equation can be written as

$$n_i^* = \frac{\mathrm{d}n_i}{\mathrm{d}t} = h_i(\pi_i - \pi_{i0}) \tag{5.5}$$

where π_{i0} is a suitable constant which may be set equal to zero and h_i is a positive constant measuring the speed of adjustment. Dreze and Sheshinski offer the following interpretation of the dynamic entry–exit process given in Equation (5.4): it assumes that new plants of type i are built whenever the profits are positive, that is $\pi_i > \pi_{i0}$; likewise the plants of type i are scrapped or not replaced when profits are negative. Since the number of firms n_i is discrete and integral, it may be more appropriate to replace it by output y_i which is proportional to n_i. Then the dynamic process reduces to

$$\dot{y}_i = \frac{\mathrm{d}y_i}{\mathrm{d}t} = h_i(\pi_i - \pi_{i0}) \tag{5.6}$$

This can be easily shown to be stable by using the Lyapunov function

$$V(y_i, y_i^*) = \frac{1}{2}(\pi_i - \pi_i^*)^2 \tag{5.7}$$

in terms of the distance between the solution path $\pi_i(t)$ and the equilibrium $\pi_i^*(t)$, where $\pi_i(y_i^*(t)) = 0$, that is there is no excess profit at equilibrium.

Since we have $\partial\pi_i(y_i(t))/\partial y_i < 0$ for any positive output y_i, where $\partial\pi_i/\partial y_i \leq 0$ for all $y_i \geq 0$, the equilibrium y_i^* associated with π_i^* is unique. Thus it holds that $\pi_i(y_i(t)) = 0$ whenever $y_i(t) = y_i^*$ and $\pi_i(y_i(t)) < 0$ for all $y_i(t)$ such that $y_i(t) > y_i^*$. Likewise we have $\pi_i(y_i(t)) > 0$ for all $y_i(t)$ such that $y_i(t) < y_i^*$, since it holds here that $p > \mathrm{MC}_i$.

Now we differentiate the Lyapunov function given in Equation (5.7) with respect to time and obtain

$$\dot{V} = \frac{\mathrm{d}V}{\mathrm{d}t} = h_i(y_i(t) - y_i^*)\ \pi_i(y_i(t)) \tag{5.8}$$

Clearly $\dot{V}$ is negative for all t. This is so because $\pi_i(y_i(t))$ is negative whenever $y_i(t) > y_i^*$ and, also $\pi_i(y_i(t))$ is positive whenever $y_i(t) < y_i^*$. Hence we have $\dot{V}(t)$ negative for all t whenever $y_i(t) \neq y_i^*$ and it is continuous. Hence we have the convergence

$$\lim_{t\to\infty} y_i(t) = y_i^* \tag{5.9}$$

Thus the entry–exit dynamic is stable in the sense of Lyapunov. In case the entry dynamic is nonlinear, that is

$$\dot{y}_i = g_i(\pi_i(y_i(t))), \quad g_i(0) = 0$$

but it holds that $\dot{V}(t)$ is negative for all t whenever $y_i(t) \neq y_i^*$, then we have local stability in the sense

$$\lim_{t\to\infty} y_i(t; y_{i0}, t_0) \to y_i^* \tag{5.10}$$

for a suitable initial value y_{i0} at time $t_0 = 0$, for example close to the equilibrium. This implies global stability if the convergence (Equation (5.10)) holds for any initial conditions (see for example Coddington and Levinson (1955)).

A third form of entry dynamics under competitive adjustment has been considered by Novshek and Sonnenschein (1986). It has three stages, where it is assumed that the industry or the market for a single homogeneous good has an inverse demand function $p(Y)$ and a continuum of infinitesimal firms $\beta \in [0, \infty]$ each with a strictly convex cost function $c(y, \beta)$. By aggregating the set of active firms one obtains the aggregate output Y. In the first stage, given any aggregate output Y, price p adjusts to the inverse demand function $p(Y)$. This is the usual tatonnement price dynamics, that is price changes in proportion to excess demand $(D - Y)$ where $D = p^{-1}(p)$ and this dynamics is stable since the inverse demand function is downward sloping. The second stage is quantity adjustment among the active firms $\beta \in [0, b],\ b < \infty$, till the short-run equilibrium output is attained. The adjustment process assumes that each active firm changes its output $y(\beta, t)$ in proportion to the difference between its desired output $y^*(\beta, t)$ at the current prices and the actual output $y(\beta, t)$:

$$\dot{y}(\beta, t) = k[y^*(b, t) - y(t)] \tag{5.11}$$

where it is assumed $k = 1$, that is instantaneous adjustment. This yields the adjustment process for aggregate output as

$$\dot{Y}(t) = Y^*(b) - Y(t) \tag{5.12}$$

Since the inverse demand function has a negative slope, it follows that whenever $Y(t) > Y^*(b)$, then $p(Y(t)) < p(Y^*)$ and hence $\dot{Y}(t) < 0$. Similarly for $Y(t) < Y^*(b)$, one gets $\dot{Y}(t) > 0$. Stability then follows from the fact that the inverse demand function cuts the supply curve from above.

The third stage consists of the long-run entry–exit dynamics, where firms' efficiencies are assumed to be ordered by β. It is assumed also that the firms with the highest incentive enter first, that is if the firm b in the order $\beta \in [0, b]$ receives strictly positive profits, then all firms with β just above b enter. Likewise for exit. The rate of entry is then assumed to be proportional to the profit at the margin, that is

$$\dot{b}(\tau) = k\pi \tag{5.13}$$

where $k = 1$ and profit is $\pi = p(Y^*(b(\tau))\, Y^*(b(\tau)) - c(Y^*(b(\tau)))$. This adjustment process is also stable, since $b(\tau) > 1$ implies that the profit of firm $b(\tau)$ is negative. Likewise $b(\tau) \leq 1$ implies that $\pi = \pi(b(\tau)) \geq 0$. Again this follows from the fact that the long-run inverse demand curve cuts the long-run supply curve from above, that is when aggregate output is higher than that at the margin, price becomes lower and this deters entry through reduced profit expectations.

5.3 Instability in competitive adjustment

So far we have discussed the competitive adjustment processes which maintain stability under free entry–exit assumptions. This stability is at least local, when the initial point is chosen suitably close to the equilibrium point, for example the linearizing approximations provide such points in the neighborhood of the equilibrium point. If local stability holds for any initial point close to the equilibrium point or not, then we have global stability.

Three important sources of instability in the competitive adjustment dynamics will be discussed here: The first arises when there is an adjustment lag or delay as in the cobweb model of agricultural goods which have fixed gestation lags. The second arises in the dual adjustment process (5.1) and (5.2) where there are increasing returns to scale. Matsuyama (1991) has analyzed similar situations where multiple equilibria can arise when there exist scale economies in the manufacturing sector for example. Finally we consider the Novshek–Sonnenschein type of adjustment dynamics but modify its longrun entry–exit dynamics where firms maximize the discounted value of future expected profits with an exogenous discount rate.

Consider the case when firms have identical cost structures, and adjustment equation is of the linear form (Equation (5.2)) with a linear marginal cost and a linear demand function. However, we assume that there is a production lag $\tau > 0$ so that the marginal cost is of the form $c(y) = c_0 + y(t - \tau)$ with $c_0 > 0$. The entry equation then takes the form

$$\begin{aligned} \dot{y}(t) &= a(\alpha - n\beta y(t) - c_0 y(t - \tau)) \\ \text{that is } \dot{y}(t) &= a\alpha - a_0 y(t) - a_1 y(t - \tau) \end{aligned} \tag{5.14}$$

where $a_0 = a\beta n$, $a_1 = ac_0$ and a, τ are positive. Clearly the entry Equation (5.14) is a linear differential equation with a deviating or retarding argument τ, that is, it is a linear mixed difference-differential equation. This type of equation has been studied by Bellman and Cooke

(1963), Elsgolts and Norkin (1973) and others. A convenient method of analyzing the stability or instability of such equations is through the D-partition method using the characteristic equation called the characteristic quasipolynomial. For the system given by Equation (5.14) the quasipolynomial takes the form

$$\phi(\lambda) = \lambda + a_0 + a_1 \, \mathrm{e}^{-\lambda\tau} \tag{5.15}$$

This has a zero root $\lambda = 0$ for

$$a_0 + a_1 = 0 \tag{5.16}$$

This straight line is one of the lines forming the boundary of the D-partition. Another boundary of the D-partition can be obtained from the purely imaginary root iz used for $\lambda = x + iz$ in $\phi(\lambda) = 0$, that is

$$iz + a_0 = a_1 \, \mathrm{e}^{-\tau iz} = 0$$

or

$$iz + a_0 + a_1(\cos \ \tau z - i \ \sin \ \tau z) = 0$$

On separating the real and imaginary parts we obtain the equation of the D-partition boundaries in parametric form as

$$a_0 + a_1 \ \cos \ \tau z = 0, \quad z - a_1 \ \sin \ \tau z = 0, \quad 0 < z < \frac{\pi}{\tau}$$

or

$$a_1 = \frac{z}{\sin \tau z}, \ a_0 = \frac{-z \cos \tau z}{\sin \tau z} \tag{5.17}$$

Various regions of stability and instability can now be determined from the areas partitioned by the boundaries given by Equations (5.16) and (5.17). Thus for $a_0 > 0$ and $a_1 = 0$ the degenerate quasi-polynomial (5.15) has no roots with positive real parts. Hence the region specified by $a_0 > 0$ and $a_1 = 0$ is the region of asymptotic stability of solutions of the equilibrium Equation (5.14). Now consider a second region for the polynomial $\phi(\lambda) = 0$ when a_1 is a fixed positive constant but a_0 varies. On taking the differential of Equation (5.15) around $\lambda = 0$ one obtains

$$\mathrm{d}\lambda = \frac{-\mathrm{d}a_0}{(1 - a_1\tau)} \tag{5.18}$$

across the straight line given in Equation (5.16). If $a_1\tau < 1$ and a_0 decreases, then $d\lambda > 0$, that is the real part of the root gets a positive increment. Since any root with a positive real part contributes to instability, we obtain an unstable equilibrium path for Equation (5.14). Also whenever $a_1\tau > 1$ and $da_0 > 0$ we obtain $d\lambda > 0$, that is a positive root, hence instability. This result can also be obtained on the boundary of the D-partition specified by Equation (5.17), that is

$$\begin{aligned} dx &= -Re\frac{da_0}{1 - a_1\tau e^{-\tau\lambda}} = -Re\frac{da_0}{1 - a_1\tau e^{-\tau iz}} \\ &= -Re\frac{da}{1 - a_1\tau(\cos\ \tau z - i\sin\ \tau z)} \\ &= \frac{(1 - a_1\tau\cos\ \tau z)\, da_0}{[(1 - a_1\tau\ \cos\ \tau z)^2 + a_1^2\tau^2\sin^2\ \tau z]} \end{aligned} \tag{5.19}$$

Clearly $\cos\ \tau z < 0$ for $a_1\tau > 1$. Hence the sign of the real part dx is opposite to the sign of da_0. Hence the pair of complex conjugate roots gain positive real parts, that is the adjustment dynamics (Equation (5.14)) exhibit instability.

Elsgolts and Norkin (1973) have defined absolute stability as follows. The solution of the mixed difference–differential equation

$$\dot{y}(t) = f(t, y(t),\ y(t - \tau_1),\ \ldots, y(t - \tau_m))$$

is called absolutely stable, if it is stable asymptotically for arbitrary, constant nonnegative lags τ_j.

It is clear from the boundaries given in Equation (5.17) specified by the purely imaginary root of the characteristic quasipolynomial that the domain represented by the inequalities

$$a_0 > 0 \text{ and } |a_1| < a_0 \tag{5.20}$$

exhibits asymptotic stability of the solution of Equation (5.14) for any $\tau \geq 0$. Hence the adjustment dynamics in this domain has absolute stability. Since a_0 involves the slope of the inverse demand function and a_1 that of supply function, the condition $a_1/a_0 < 1$ requires that demand adjusts faster than the supply in order to ensure absolute stability. Outside the domain specified by Equation (5.20) we have regions of instability and also oscillations when the roots of the quasi-polynomial are complex. For fixed $a_1 > 0$ and varying a_0 we can state the instability result as follows.

THEOREM 1. *If the quasi-polynomial $\phi(\lambda)=0$ in Equation (5.15) has a positive real part of $d\lambda$ for any variation da_0 and $\tau>0$, then the adjustment dynamics given in Equation (5.14) is asymptotically unstable. If any root is complex, then it exhibits undamped oscillations.*

PROOF. The characteristic quasipolynomial $\phi(\lambda)=0$ has a zero root when Equation (5.16) holds. The straight line generated by Equation (5.16) yields one boundary of the D-partition. On keeping a_1 fixed at $a_1>0$, we increase a_0 so that $da_0>0$. If $a_1\tau>1$ then it follows from Equation (5.18) that $d\lambda>0$ around $\lambda=0$. Hence there exists a domain in the D-partition space of characteristic roots at which there exist roots with positive real parts. Hence in this domain the adjustment dynamics given in Equation (5.14) is asymptotically unstable, and with complex roots the oscillations are not damped as $t\to\infty$.

Several comments on this result are in order. First of all, if the inverse demand function $p=p(y)$ and the marginal cost function $c=c(y)$ are nonlinear, then the nonlinear adjustment Equation (5.2) can be linearly approximated and the homogeneous part can be written as

$$\begin{aligned}\dot{y}(t)&=a(p'y(t)-c'y(t-\tau))\\&=a_0y(t)-a_1y(t-\tau)\end{aligned}\tag{5.21}$$

where $a_0=ap'$, $a_1=ac'$ and prime denotes partial derivative. Since p' is negative, one obtains the mixed differential–difference equation in the form of Equation (5.14). Clearly if the solution of Equation (5.21) is asymptotically stable for any arbitrary positive large τ_j, then it is asymptotically stable. Furthermore, if this type of stability holds for arbitrary initial conditions, then it is globally and absolutely stable.

Secondly, we have

$$\dot{y}(t)+(a_0+a_1)\,y(t)=0\tag{5.22}$$

We may compare the solution regions of (5.21) and (5.22).

Equation (5.22) has no lags and the stability region for the solution of this ordinary linear differential equation is defined by the inequality $(a_0+a_1)>0$. On comparing the regions of solutions of Equations (5.21) and (5.22) one may easily derive the following results:

(a) If the solutions of Equation (5.22) are unstable, then the solutions of Equation (5.21) are also unstable for any $\tau>0$. This follows from the fact that for sufficiently small τ, the Equation (5.21) closely approximates Equation (5.22). For example, if $a_0<0$ but $|a_0|>|a_1|$ the

Equation (5.22) is unstable. Under these conditions the characteristic quasipolynomial has at least one root with a positive real part and hence it is unstable.

(b) If the solutions of Equation (5.22) are asymptotically stable and $a_0 > a_1 > 0$, then the solutions of Equation (5.21) are also asymptotically stable for any $\tau > 0$. □

Note that if we consider the Walrasian adjustment dynamics in the excess demand form (Equation (5.1)) and assume linear demand and supply functions, where supply adjusts with a lag, then we obtain

$$\dot{p}(t) = b[d_0 - d_1 p(t) - (s_0 + s_1 p(t-\tau))]$$

or,

$$\dot{p}(t) = (bd_0 - bs_0) - bd_1 p(t) - bs_1 p(t-\tau) \tag{5.23}$$

If d_1, s_1 are positive we obtain the system

$$\dot{p}(t) = \beta - \beta_0 p(t) - \beta_1 p(t-\tau)$$

where $\beta = bd_0 - bs_0$, $\beta_0 = bd_1$, $\beta_1 = bs_1$. The difference–differential equation (Equation (5.23)) is of the same form as Equation (5.14) and hence there exist domains of both asymptotic stability and instability as analyzed before.

Next we consider the dual system given by Equations (5.1) and (5.2) as the Walrasian adjustment dynamics and consider increasing returns as the source of instability. Here p^* and y^* are defined as the equilibrium price and output, respectively, which make $p^* = c(y^*)$ and $d(p^*) = y^*$. It is assumed that (p^*, y^*) are unique. If marginal cost $c(y)$ is constant or increasing (i.e., $c' = \partial c/\partial y > 0$) and the demand curve is always downward sloping (i.e., $d' = \partial d/\partial p$) then one can show that the point (p^*, y^*) is a globally stable equilibrium. This is so because by linearizing the system given by Equations (5.1) and (5.2) one obtains the homogeneous system:

$$\begin{pmatrix} \dot{y} \\ \dot{p} \end{pmatrix} = \begin{bmatrix} -ac' & a \\ -b & bd' \end{bmatrix} \begin{pmatrix} y \\ p \end{pmatrix}$$

which has the characteristic equation

$$\phi(\lambda) = \lambda^2 + (ac' - bd')\,\lambda + ab(1 - c'd') = 0 \tag{5.24}$$

This quadratic Equation (5.24) has two roots with negative real parts if and only if

$$ac' - bd' > 0 \quad \text{and} \quad ab(1 - c'd') > 0 \tag{5.25}$$

and the roots are complex if

$$(ac' - bd')^2 < 4ab(1 - c'd') \tag{5.26}$$

Clearly Equation (5.25) holds if $c' > 0$, $d' < 0$ and a, b are positive. Hence $p(t)$ and $y(t)$ converge to their respective equilibrium p^* and y^* as $t \to \infty$.

However, if there is increasing returns and hence $c' < 0$ we may have the situation

$$c' < \frac{1}{d'} < 0 \tag{5.27}$$

when the returns to scale are sufficiently large. In this case $c'd' > 1$ and hence there are two solutions of Equation (5.24), one positive and one negative. Hence (p^*, y^*) is now a saddle point. This implies that there is a stable manifold along which convergence is purely towards (p^*, y^*) and an unstable manifold along which motion diverges away from (p^*, y^*), that is one is a stable path, the other an unstable one. Hence we can state the following result:

THEOREM 2. *Whenever the increasing returns is very large so that* $c' < 1/d' < 0$ *holds, then the dual adjustment dynamics defined by Equations (5.1) and (5.2) have an unstable manifold which exhibits asymptotic instability. For* $-c' > -d' > 0$, *if the roots are complex then the unstable manifold exhibits cyclical fluctuations or oscillations.*

PROOF. The condition $c' < 1/d' < 0$ implies that $c'd' > 1$. Hence the two roots are given by

$$\lambda_1 = \theta, \lambda_2 = -(ac' - bd') - \theta, \theta > 0$$

where

$$[(ac' - bd')^2 - 4ab(1 - c'd')] = ac' - bd' + 2\theta, \ \theta > 0$$

Whenever the adjustment coefficients a, b are equal and $-c' > d' > 0$, then the roots have a positive real part. The roots are complex under this condition if

$$(c' + d')^2 < 4, \quad a = b \text{ and } c'd' < 1$$

Hence the solution exhibits explosive oscillations.

Two comments are in order. First, the adjustment Equation (5.2) may be modified as

$$\dot{y} = a\tilde{\pi}, \quad \tilde{\pi} = p - \text{AC}$$

where π is profitability per unit of output and AC is average cost at a minimal level. Second, it is assumed there that in the long run the markets clear when $p^* = c(y^*)$, $d(p^*) = y^*$. But in the real world this may not happen, that is markets do not clear, so that demand and outputs are not equal, implying that there is either continuing accumulation or decumulation of inventories, or continuing rationing. Such disequilibrium has been analyzed by Benassy (1982) and others.

Finally, we consider the entry–exit dynamics where long-run returns are explicitly introduced and the entry–exit equation is written as

$$\dot{y}(t) = h(R(y) - C(y)), \quad h > 0 \tag{5.28}$$

where profit π is long-run returns $R(y)$ minus long-run costs $C(y)$. Long-run return at time t is the discounted value of short-run return $R = \int_t^\infty r(y)e^{-\rho(\tau - t)}d\tau$. Differentiation of R yields

$$\dot{R} = \rho R - r(y) \tag{5.29}$$

where ρ is the exogenous rate of discount assumed to be positive.

The discounted cash flow of returns in the entry Equation (5.28) is more appropriate in situations where future prices and demands are unknown and only their expected values can be used. Thus $R(y)$ can be viewed as the expected market value of returns of the incumbent firm. In a static equilibrium a free entry condition would imply that profit is zero, that is $R(y) = C(y)$ and the adjustment Equation (5.28) assumes that firms enter the industry (measured by either $\dot{n}$ or $\dot{y}$) when $R(y) > C(y)$ and exit when profit $\pi = R(y) - C(y)$ is negative.

For analyzing the competitive adjustment dynamics specified by Equations (5.28) and (5.29), Equation (5.28) is linearized and the characteristic equation $\phi(\lambda) = 0$ is specified for the homogeneous part as before. This yields

$$\lambda^2 - (\rho - hC')\lambda + hr' = 0 \tag{5.30}$$

□

THEOREM 3. *Whenever it holds that $\rho > hC'$ and $f' > 0$, then the adjustment dynamics defined by Equations (5.28) and (5.29) have an unstable manifold, which exhibits asymptotic instability. Furthermore if $\rho > hc'$ along with $(\rho - hC')^2 < 4hr'$, $r' > 0$, then the unstable manifold exhibits cyclical fluctuations or oscillations.*

PROOF. The characteristic roots of Equation (5.30) are

$$\lambda_1 = \theta, \quad \lambda_2 = \rho - hC' - \theta, \quad \theta > 0 \text{ if } (\rho - hC')^2 > 4hr', \quad f' > 0$$

hence at least one root $\lambda_1 = \theta$ is positive implying an unstable manifold. If $\rho - RC' > \theta$, $\theta > 0$ then both roots are unstable. Hence if $\lambda_1 = \theta$, $\theta > 0$, but $\lambda_2 < 0$, then we have a saddle point, having one stable and another unstable manifold.

$$\text{If } (\rho - hC')^2 < 4hr', \quad r' > 0$$

then the roots are complex and if $\rho > hC'$ then we have an unstable manifold with cyclical fluctuations or oscillations. Along this manifold the Walrasian entry/exit dynamics is asymptotically unstable and oscillatory.

Two comments are in order. First, we may consider the entry dynamics (Equation (5.28)) in terms of unit revenues $q(y)$ and costs $c(y)$, that is

$$q = q(y) = \int_t^\infty p(y)\mathrm{e}^{-\rho(\tau - t)}\mathrm{d}\tau$$

$$c = \text{long-run average cost}$$

then the adjustment dynamics simplifies to

$$\dot{y}(t) = h(q - c)$$

$$\dot{q} = \rho q(y) - p(y)$$

The unstable manifold is then specified by $\rho > hc' > 0$ and $p' < 0$, and the roots are $\lambda_1 = \rho - hc' + \theta$, $\lambda_2 = -\theta$, $\theta > 0$. Clearly this specifies a saddle point. Secondly, the entry–exit equations (Equations (5.28) and (5.29)) can also be specified in terms of the number (n) of firms, that is

$$\dot{n}(t) = h(R(n) - C(n))$$

$$\dot{R} = \rho R(n) - r(n)$$

Thus the unstable manifold can be specified in terms of the divergence of number of firms, where the equilibrium is specified by n^* at which $R(n^2) - C(n^*)$, $n^* > 0$, that is, at this level of n^* there is no new entry into (or exits from) the industry in the long run. □

5.4 Adjustment with entry barriers

The assumption of free entry–exit is critical to the competitive selection of firms and Walrasian adjustment. It implies the absence of any potential and actual entry barriers and hence zero cost of entry. Even in oligopolistic models with Cournot–Nash equilibrium, Novshek (1980) has shown the validity of the folk theorem which says that if the oligopolistic firms selling a homogeneous product are small relative to the total market, then the market outcome is approximately competitive, that is, the competitive entry model of Walrasian adjustment specified in Equations (5.1) and (5.2) holds approximately.

When the conditions of the folk theorem fail to hold, so that firms are not small relative to the total market, the competitive entry dynamics have to be modified. The entry-blocking strategies and/or barriers to entry have to be explicitly introduced. Thus on defining net entry (N) by the difference of gross entry over exit, the entry equation would now depend on BTE variables, excess profits and new technology which may shift the demand and cost frontiers, that is

$$\dot{N} = \frac{\mathrm{d}n}{\mathrm{d}t} = f(\mathrm{BTE},\ \pi - \pi^0,\ \tau) \tag{5.31}$$

where π^0 is the average industry level profits and τ is a shift parameter. Thus if BTE is very significant it retards new entry $(\dot{N} < 0)$. Excess profits $\pi > \pi^0$ over the industry average tend to invite positive entry $(\dot{N} > 0)$, and new innovations by outside firms indexed by τ may reduce costs and thereby increase expected profitability and hence positive entry by outside firms. The empirical and econometric applications of the dynamic entry equation (Equation (5.31)) by Mata (1995) and others have considered BTE in several forms, for example existence of patents, largeness of the MSE, largeness of sunk cost and significant product differentiation. Statistically significant results are obtained for the MSE variable and large investment in sunk cost. Empirical studies by Sengupta (2004) for modern technology-intensive industries such as computers and microelectronics have shown a very significant role for R&D investment variables in firm growth. Large R&D investment and innovations may act as a potential barrier to entry, since it may reduce long-run average costs. We consider here two types of entry barriers.

One is represented by large capital investment (k) and the other the R&D investment (u) generating economies of scale. In the first case we assume total costs to be inclusive of both production cost and entry cost, where the latter depends on k. In the second case we consider the cost of R&D investment separately from production cost and the cost of R&D investment is assumed to imply the existence of entry barrier or deterrence through high entry cost.

In the first case we assume that the selection process of firms in the evolution of industry follows a Cournot–Nash equilibrium, when firms operate in an oligopolistic market with a single homogeneous output. Suzumura (1995) considered such a formulation in two stages. The first stage determines the equilibrium solution of each incumbent firm with profits $\pi(n)$ and output $y(n)$ dependent on the number of firms n, where

$$\pi(n) = y(n)\, p - C(y(n)) \tag{5.32}$$

where $p = p(Y)$ is price and $C(\cdot)$ is total production cost. The second stage introduces the entry dynamics in the long-run equilibrium as

$$\dot{n} = \frac{\mathrm{d}n}{\mathrm{d}t} = a\pi(n), \quad a > 0 \tag{5.33}$$

where the positive constant a denotes the adjustment coefficient. The entry dynamics assumes that if $\pi(n) > 0$ (or < 0), potential competitors (or incumbent firms) will be induced to enter into (or exit from) this profitable (or unprofitable) industry. Suzumura defines the long-run equilibrium number of firms as the stationary point (n^*) of (5.33), that is

$$\pi(n^*) = 0 \tag{5.34}$$

If the inverse demand function $p(Y)$, $Y = \sum_{i=1}^{n} y_i$, has a negative slope, then it can be shown that the long-run equilibrium number (n^*) of firms is unique and also stable as $t \to \infty$. To prove stability he used the Lyapunov function

$$V(t) = \left(\frac{1}{2}\right)[n(t) - n^*]^2$$

to derive the condition

$$\dot{V} = \frac{\mathrm{d}V(t)}{\mathrm{d}t} = a(n(t) - n^*)\pi(n, n_0) \tag{5.35}$$

where n_0 is the initial number of firms assumed given. Clearly if $n(t) > n^*$, then $\pi(n) < 0$ since $\mathrm{d}\pi(n)/\mathrm{d}n < 0$. Also, if $n(t) < n^*$, then $\pi(n) > 0$.

Both cases yield a negative value for $\dot{V}$. Furthermore if $n(t) = n^*$, then $\dot{V} = 0$. Hence the convergence $\lim_{t \to \infty} n(t) = n^*$ is asymptotically stable. It is globally stable if it holds for arbitrary initial conditions.

We generalize this formulation in several directions. First, we combine entry costs and production costs, where entry costs are associated with entry barriers. Secondly, output is used instead of the number of firms in the entry–exit dynamics. Finally, the equality of marginal revenue and marginal cost implicit in the first stage of Cournot–Nash equilibrium formulation of Suzumura's model is used in a dynamic formulation as follows:

$$\dot{y}(t) = h(M(y) - c(y, k)) \tag{5.36}$$

$$M(y) = \sum_{t}^{\infty} \mathrm{e}^{-\rho(\tau - t)} m(y) \mathrm{d}\tau \tag{5.37}$$

where $M(y)$ is the discounted value of future marginal revenue $m(y)$ with a given discount rate $\rho > 0$, and $c(y, k)$ is the combined marginal cost of entry and production, with k denoting large capital investment. Thus the net present value of each entry into the industry is the difference between the long-run marginal revenue and marginal costs. In a state of static equilibrium we have $M(y) = c(y, k)$, implying zero-profits $\pi(y, k) = 0$ as in the Cournot–Nash equilibrium (Equation (5.34)) of the Suzumura model. On differentiating (5.37) one obtains

$$\dot{M} = \frac{\mathrm{d}M(y)}{\mathrm{d}t} = \rho M(y) - m(y) \tag{5.38}$$

Thus we have the long-run dynamics defined by Equations (5.36) and (5.38) specifying the industry evolution path in a Cournot–Nash framework. When the entry barrier represented by k affects the MSE level of output, we may express costs $c(y, k)$ as $c(y)$, and our dynamic system of industry evolution becomes

$$\dot{y}(t) = h(M(y) - c(y))$$

and

$$\dot{M}(y) = \rho M(y) - m(y) \tag{5.39}$$

where h is a positive constant denoting the speed of adjustment. On combining the two Equations of (5.39) and linearizing around the equilibrium point y^* at which optimal profits is zero, one obtains the linear homogenous system

$$\ddot{y} = h(\rho M' y - m\dot{y}) - hc'\dot{y}$$

This yields the characteristic equation

$$\lambda^2 + hc'\lambda - (h\rho M' - m') = 0 \tag{5.40}$$

Hence we can state the following result.

THEOREM 4. *Whenever the two roots of (5.40) are real and it holds that $m' - h\rho M' < 0$ then the adjustment dynamics have unstable manifold with the equilibrium output y^* being a saddle point. Furthermore, if the roots are complex and marginal cost is declining, the unstable manifold exhibits cyclical fluctuations or oscillations.*

PROOF. The two real roots of the Equation (5.40) may be derived as

$$\lambda_1, \lambda_2 = \frac{1}{2}[-hc' \pm (hc' + 2\theta)] = \theta, \quad -(hc' + \theta), \theta > 0$$

Hence one root is positive if $\theta > hc'$. Both roots are positive if $-hc' > 0$, where $c' < 0$. Hence there exists at least one unstable manifold.

If the roots are complex and $c' < 0$, then the two roots have positive real parts. Hence the equilibrium point y^* exhibits cyclical fluctuations.

If both roots are real and positive, then we have an unstable node. Also when the roots are complex-valued and $c' < 0$, then the stationary state y^* is an unstable focus. □

Two comments are in order. First, declining marginal costs ($c' < 0$) tend to impart instability through the positive roots. Hence if $c' > 0$ due to increasing entry barriers, the stationary point y^* is likely to be stable. Secondly, if ρ is very small, so that $h\rho M'$ can be neglected, then the roots would exhibit divergence if $c' < 0$ and $m' \leq 0$.

Now we consider the entry costs separate from production cost and we assume that the potential entry cost is due to the high investment in R&D by the incumbent firms. We denote entry cost by z and assume it to be fixed. We write the entry equation as

$$\dot{y} = \alpha(v - z), \quad \alpha > 0 \tag{5.41}$$

where v is the long-run discounted value of the incumbent firm's profit $\pi(y)$:

$$v = v(y) = \int_t^\infty \pi(y) e^{-\rho(\tau - t)} dx \tag{5.42}$$

where ρ is the positive discount rate assumed to be exogenous. The net present value of each entry into the industry is the difference between the market value of the incumbent firm and the opportunity cost, that is the firm's value in the best alternative industry. On differentiating Equation (5.42) one obtains

$$\dot{v} = \rho v - \pi(y) \tag{5.43}$$

Thus the two dynamic Equations (5.41) and (5.43) constitute the adjustment process with entry barriers represented by the cost of entry z. When the entry cost is not fixed but depends on R&D expenditure u, profit π would also depend on u and hence the dynamic adjustment system would be

$$\begin{aligned} \dot{y} &= \alpha(v - z(y, u)) \\ \dot{v} &= \rho v - \pi(y, u) \end{aligned} \tag{5.44}$$

Note that the positive profits $\pi = \pi(y, u) > 0$ alone are not sufficient to induce positive entry, since the entry costs $z(y, u)$ may still deter entry, that is $z(y, u) > v$. Also the entry and exit are endogenously determined at each moment of time t. In a static model the equilibrium number of firms and hence output would be at the level $v = z$. Thus entry (exit) occurs in the form of $\dot{y} > 0$ (or $\dot{y} < 0$) whenever $v > z$ (or $v < z$).

This is a two-stage model, where positive (or negative) profits induce (or deter) entry in the first stage, but actual entry (or exit) occurs if and only if the net present value (v) of entry (or exit) exceeds (or falls short of) the cost of entry (z).

The R&D parameter u is the effort made by the incumbent firm in product innovation at time t. We now assume that for each output y the profit minus entry cost $\pi(y, u) - z(y, u)$ has an interior maximum in u denoted by $u^* = u^*(y)$, that is

$$u^*(y) = \max_u \ \pi(y, u) - z(y, u) \tag{5.45}$$

Clearly this assumption reflects decreasing returns to R&D at the firm level. On using these values u^* in Equation (5.44) one obtains the dynamic system

$$\begin{aligned} \dot{y} &= \alpha(v - z(y)) \\ \dot{v} &= \rho v - \pi(y) \end{aligned} \tag{5.46}$$

On linearizing this system around y^* at which $v = z(y)$ one can easily derive the homogenous linear system

$$\begin{pmatrix} \dot{y} \\ \dot{v} \end{pmatrix} = \begin{bmatrix} -\alpha z' & \alpha \\ -\pi' & \rho \end{bmatrix} \begin{pmatrix} y \\ v \end{pmatrix} \tag{5.47}$$

This implies the following result on instability:

THEOREM 5. *Whenever ρ exceeds $\alpha z'$ there exists an instable manifold of the equilibrium. When entry cost is fixed so that $z' = 0$ and $\pi' > 0$ with a large α, then the unstable manifold exhibits cyclical fluctuations or oscillations.*

PROOF. The two roots of the characteristic equation based on Equation (5.47) may be easily derived as

$$\lambda_1, \lambda_2 = \frac{1}{2}[(\rho - \alpha z') \pm ((\rho - \alpha z')^2 - 4\alpha(\pi' - \rho z'))^{1/2}]$$

If the roots are real, then they can be expressed as

$$\lambda_1 = \theta, \quad \lambda_2 = -(\theta + \alpha z' - \rho), \quad \theta > 0 \text{ if } \pi' < \rho z'$$

where at least one root is positive if $\rho > \alpha z'$. If $z' = 0$ this is necessarily so. The saddle point exists when $\theta + \alpha z' > \rho > 0$, since the two roots are with opposite signs. In case $z' = 0$, that is constant entry cost but $\pi' = \pi'(y^*) > 0$ and α is sufficiently small, both characteristic roots are positive and the equilibrium is an unstable node. If $z' = 0$ but $\pi' > 0$ with α large, then the roots are seen to be complex valued with positive real parts and hence the stationary state y^* is an unstable focus with cyclical fluctuations or oscillations. □

Some implications of the above instability results may be noted. First of all, there may exist other sources of instability when there is no interior maximum in R&D effort as assumed in Equation (5.45), that is three may be local increasing returns to R&D. Folster and Trofimov (1997) have explored the implications of an S-shaped profit function in terms of empirical data of Swedish firms during 1988 and 1990. Their statistical estimates support the notion that there can be ranges of aggregate increasing returns to research in which an industry can converge to different equilibria, some stable and some unstable. Secondly, the case of cyclical dynamics in the presence of complex roots with positive real parts may imply a multiplicity of equilibrium paths. Since linearizing around an equilibrium point implies a local analysis, the multiplicity of equilibria in the global nonlinear case presents another source of instability. Finally, R&D investment yields externalities in the sense that knowledge acquired in one firm spills over to other firms and this type of spread of knowledge may lead to aggregate increasing returns to scale for the industry. Such externalities would modify both entry costs z and profits π thus affecting the stability characteristics of the dynamic adjustment equations (Equation (5.44)).

5.5 Potential entry and preemptive behavior

A major incentive for setting entry barriers is to deter potential entry. Gaskins (1971) developed the limit pricing model of a dominant or monopoly firm based on the assumption that a positive entry occurs whenever the dominant firm charges a price much higher than the limit price. He measured entry in terms of the rate of increase in output of the new entrants and assumed it to be proportional to the excess of price over the limit price. This type of model has been extended by Dixit (1980), Gilbert (1986) and others in terms of an entry-deterring output level, which may be called the limit output. This formulation assumes that firms have access to the same technology as the monopoly or dominant firm and can produce any output level y_j at a constant marginal cost c, provided the output is at least $\bar{k}$, that is $\bar{k}$ is an MES of output for any firm. The output of the incumbent firm is y_0 where the industry demand is assumed to be linear, that is $p(Y) = a - bY$. The new entrant takes y_0 as given and enters if the resulting price is such that

$$p(y_0 + y_1) > c \text{ for any } y_1 \geq \bar{k} \tag{5.48}$$

This yields a critical entry-deterring output

$$\overline{Y} = \frac{(a-c)}{b-\bar{k}}$$

where the incumbent may either deter the entrant by setting $y_0 = \overline{Y}$ or it may allow entry. If the monopoly firm's output is less than the limit output $\overline{Y}$, then the entry is not blockaded by the profit-maximizing output of the established firm. Thus the incumbent firm's profit with entry deterrence is

$$\bar{\pi}_0 = \overline{Y}(p(\overline{Y}) - c) \tag{5.49}$$

But if the incumbent allows entry, it should choose output y_0 after taking into account the output of the entrant firm. Thus the incumbent firm chooses y_0 optimally with foresight by maximizing the profit

$$\pi_0(1) = \max_{y_0} y_0[p(y_0 + R_1) - c] \tag{5.50}$$

where $R_1 = R_1(y_0)$ is the output reaction function of firm 1 determined by the solution to the following problem:

$$\pi_1(1) = \max_{y_1} y_1[p(y_0 + y_1) - c]$$

Thus it is clear that the incumbent should choose the entry-preventing output if

$$\bar{\pi}_0 \geq \pi_0(1) \tag{5.51}$$

where $\bar{\pi}_0$ and $\pi_0(1)$ are given by Equations (5.49) and (5.50) respectively. In terms of output the condition given in Equation (5.51) for entry prevention becomes

$$\overline{Y} \leq (a - c)\frac{(1 + \sqrt{\frac{1}{2}})}{2b} \tag{5.52}$$

If this limit quantity of output $\overline{Y}$ exceeds the critical level $0.85Y_c$, where Y_c is competitive output, then the established firm is better off allowing new entry. In that case the profits lost from competition are less than the profits lost by choosing a limit pricing strategy. Thus if the limit output $\overline{Y}$ is greater than $0.85Y_c = (a - c)(1 + \sqrt{0.5})/2b$, the incumbent should allow entry but if it is less then the incumbent should produce at the entry-preventing output by practicing limit pricing. Thus the dynamic adjustment path can be modeled as

$$\dot{Y} = \begin{cases} h(\overline{Y} - 0.85Y_c), \text{ if } \overline{Y} > 0.85Y_c \\ 0, \text{ if } \overline{Y} \leq 0.85Y_c \end{cases}$$

In terms of Lyapunov function theory this dynamic adjustment process can be shown to be stable.

Note that this model can be easily generalized to the case where there are many entrants and all firms act as Cournot–Nash competitors. The established firms would always try to deter entry so long as it is profitable, but there are two situations where they may not be able to prevent it. One is the case where the MES for entry is very small relative to the capacity of the established firms. In this case the preemptive investment to deter entry will not be profitable. Secondly, if the incumbent firms hold excess capacity, additional investment would not affect output and hence the preemptive investment would act as an empty threat against entry.

Thus the incumbent firms' entry deterrence strategies may have to be examined in a dynamic context much more critically. This is particularly so when R&D investment is concerned, since it generates externalities.

5.6 Competitive and noncompetitive behavior

The two types of market structures, competitive and game-theoretic, have been shown to have different implications of the entry–exit behavior. In many ways the game-theoretic markets would exhibit an adjustment behavior very similar to the Walrasian process if the oligopolistic firms are small relative to the total market and the conditions of free entry prevail. Under entry barriers the adjustment process would tend to have significant time lags where the mixed difference–differential equations would hold. One may compare the relative stickiness of prices in such markets. Two other implications may be noted. First, the presence of nonexplosive cyclical fluctuations in case of nonlinear industrial evolution would show a multiplicity of equilibrium paths. Secondly, the externality effect of R&D costs may increase the cost of risk associated with the adjustment process. This implication may be directly compared with Sutton (1998) who showed that the R&D intensity is positively correlated with the concentration in the size distribution of firms in a Cournot–Nash market structure. He measured the "toughness of competition" in such markets by a ratio α which tells us the extent to which an industry consisting of many small firms can be destabilized by a firm that outspends its many small rivals. If α is positive this means that the deviant firm that offers a higher-quality product can attain a final-stage profit much higher than the industry average.

It may be argued that a monopolistic market structure is more appropriate here than a Cournot–Nash framework. But we have not discussed here the case of differentiated products and a regime of quality ladders. Moreover the Cournot–Nash model is more appropriate as a non-Walrasian adjustment process as Sutton (1998) and others have shown.

5.7 Concluding remarks

This chapter discusses the dynamic stability characteristics of the entry and exit process in competitive and Cournot–Nash market structures. The implications of three major sources of instability such as adjustment lag, increasing returns to scale and the presence of entry barriers are considered in some detail and their instability characteristics analyzed. For the Cournot–Nash market structures the implications of toughness of competition due to large capital investment and R&D externalities are analyzed in terms of their cyclical fluctuations. However, if the incumbent firm can precommit to the level of output, this may sometimes result in an adjustment process to be stable, since the process contains both entry deterrence and accommodation.

References

Arrow, K.L. and L. Hurwicz (1962) Decentralization and computation in resource allocation, in *Essays in Economics and Econometrics in honor of Harold Hotelling* (Chapel Hill: University of North Carolina Press).

Bain, J.S. (1956) *Barriers to New Competition* (Cambridge: Harvard University Press).

Bellman, R. and K.L. Cooke (1963) *Differential–Difference Equations* (New York: Academic Press).

Benassy, J. (1982) *The Economics of Market Disequilibrium* (New York: Academic Press).

Coddington, E. and N. Levinson (1955) *Theory of Ordinary Differential Equations* (New York: McGraw-Hill).

D'Aveni, R.A. (1994) *Hypercompetition: Managing the Dynamics of Strategic Maneuvering* (New York: Free Press).

Dixit, A.K. (1980) The role of investment in entry deterrence, *Economic Journal* 90, 95–106.

Dreze, J. and E. Sheshinski (1984) On industry equilibrium under uncertainty, *Journal of Economic Theory* 33, 88–97.

Elsgolts, L.E. and S.B. Norkin (1973) *Introduction to the Theory and Application of Differential Equations with Deviating Arguments* (New York: Academic Press).

Gaskins, D.W. (1971) Dynamic limit pricing under threat of entry, *Journal of Economic Theory* 3, 306–322.

Gilbert, R.J. (1986) Pre-emptive competition, in J. Stiglitz and G. Mathewson (eds) *New Developments in the Analysis of Market Structure* (Cambridge: MIT Press).

Heal, G. (1986) Macrodynamics and returns to scale, *Economic Journal* 96, 191–198.

Judd, K. and B. Petersen (1986) Dynamic limit pricing and internal finance, *Journal of Economic Theory* 39, 368–399.

Klepper, S. (1996) Exit, entry, growth and innovation over the product life-cycle, *American Economic Review* 86, 562–583.

Mata, J. (1995) Sunk costs and the dynamics of entry in Portuguese manufacturing, in A. Witteloostuijn (ed.) *Market Evolution: Competition and Cooperation* (Dordrecht: Kluwer Academic Publishers).

Matsuyama, K. (1991) Increasing returns, industrialization and indeterminacy of equilibrium, *Quarterly Journal of Economics* 106, 617–650.

Morishima, M. (1964) *Equilibrium, Stability and Growth* (Oxford: Clarendon Press).

Novshek, W. (1980) Cournot equilibrium with free entry, *Review of Economic Studies* 47, 473–486.

Novshek, W. and H. Sonnenschein (1986) Quantity adjustment in an Arrow–Debreu–McKenzie type model, in J. Gabszewicz (ed.) *Models of Economic Dynamics* (Berlin: Springer Verlag).

Sengupta, J.K. (2003) Competition and efficiency in industry equilibrium, *Keio Economic Studies* 40, 59–73.

Sengupta, J.K. (2004) *Competition and Growth: Innovations and Selection in Industry Evolution* (New York: Palgrave Macmillan).

Sutton, J. (1998) *Technology and Market Structure* (Cambridge: MIT Press).

Suzumura, K. (1995) *Competition, Commitment and Welfare* (Oxford: Clarendon Press).

6
Model of Industry Evolution under Innovations

6.1 Introduction

How does an industry grow? What are the economic forces behind this evolution? These questions are important today in the framework of modern industrial growth, where advances in technology and innovations through R&D investments are playing very active roles. We develop here a model of industry evolution which incorporates the innovations flow and determines its growth pattern. This dynamic model characterizes the highly competitive process of adjustment in the market place, which is reflected in both output quantities and prices. Innovations add stochasticity to the competitive adjustment process.

Competition has been most intense in recent times in some of the modern high-tech industries such as computers, telecommunications and electronics. Declining prices and costs, rising world demand, increasing innovations and new product development have intensified the competitive pressure in these industries. Following Schumpeter's dynamic innovation approach, D'Aveni (1994) has characterized this state as hypercompetition. He argues that this hypercompetitive world resembles in many ways the Darwinian world of survival of the fittest, where the rival competitors get crushed if they are not on the leading edge of the innovation efficiency frontier. Sengupta (2004) has developed a two-stage model of hypercompetition, where in the first stage the firms on the efficiency frontier are identified by a nonparametric cost-efficiency model and then in the second stage the industry evolution is analyzed in terms of increased market share of the efficient firms.

Our objective here is twofold. One is to develop a dynamic model of industry evolution, which is based on the principle that increased profits will tend to invite more entry, and the barriers to entry will tend to discourage new entry. The second object is to model the impact of innovations in the form of R&D investments, which tend to augment the scale efficiency and reduce average costs at the optimal scale. This is followed by an empirical application in the computer industry. This application measures the degree of scale economies present in this industry.

6.2 Model of industry evolution

We consider an industry consisting of homogenous firms competing in R&D investments. The expected net cost of a representative incumbent firm is assumed to be a function of the number of firms n in the industry and of the R&D expenditure u, that is, $c(n, u)$. The number of firms n is assumed to be known to all incumbent firms at the beginning of time period $(t,\ t+\mathrm{d}t)$. The R&D parameter u is the effort made by the firm in product innovation at time t.

We make two specific assumptions for obtaining specific results.

Assumption 1 For each positive n the long-run cost function $c(n, u)$ has an interior minimum in u. This yields $c(n) = \arg\ \underset{u}{\mathrm{Min}}\, c(n, u)$. This assumption reflects diminishing returns to R&D at the firm level.

Assumption 2 New entry ($\dot{n} = \mathrm{d}n/\mathrm{d}t$) occurs whenever profits $\pi(n) = p - c(n)$ are positive. This assumption reflects the incentives behind new entry, that is exit occurs when profits are negative. For simplicity we assume that the entry and exit processes are linear functions of expected profits $\pi(n)$.

Since n is discrete, we replace it by output y, so that the entry–exit dynamics can be modeled as

$$\dot{y} = a(p - c(y)), \quad a > 0 \tag{6.1}$$

In a static model a free-entry condition would simply imply that the output in equilibrium is such that $p = c(y)$. This yields the equibrium number of firms in the industry.

Each incumbent firm in this competitive framework is a price taker and therefore minimizes the initial present value of long-run costs $C_0 = \int_0^\infty \exp(-rt)c(y, u)\mathrm{d}t$ in order to stay in the industry indefinitely. At any time $t > 0$ the incumbent firm's average cost is given by

$$C(t) = \int_t^\infty \exp[-r(\tau)]c(y, u)\ \mathrm{d}t$$

Differentiating this cost function one obtains

$$\dot{C} = rC - c(y, u) \tag{6.2}$$

Equations (6.1) and (6.2) specify the dynamic model of industry evolution under competitive conditions. This is called the cost model. Here the dynamic decision problem for the incumbent firm is to choose the time path of the R&D investments $u = u(t)$ which minimize the initial cost $C_0 = C(0)$ subject to the state Equations (6.1) and (6.2).

It is useful to consider two other alternative formulations of the entry–exit dynamics. The Walrasian adjustment dynamics combine Equation (6.1) with Equation (6.2) as follows:

$$\dot{p} = b(D(p) - y) \tag{6.3}$$

where market demand is $D(p)$. Since the competitive market is assumed to clear at the equilibrium price, we get the equilibrium price (p^*) and quantity (y^*) as $p^* = c(y^*)$; $D(p^*) = y^*$. Here excess demand $[D(p) - y]$ is assumed to raise the price, whenever output hits the capacity ceiling ($\bar{y}$). So long as $y < \bar{y}$ the prices are assumed not to increase. Equations (6.1) and (6.3) comprise the demand model.

Another formulation is to replace Equation (6.2) by the investment equation

$$\dot{k} = I - \delta k, \quad \delta > 0 \tag{6.4}$$

where I is gross investment, δ is the depreciation rate and k is the net stock of capital. When R&D expenditures are viewed as knowledge capital, Equation (6.4) specifies the equilibrium growth of knowledge capital or investment. Then the average cost $c(y)$ may be rewritten as $c(k, y)$ in order to reflect the cost-reducing impact of higher knowledge capital in the form of R&D inputs. When I increases over time it increases k and thereby the average cost $c(k, y)$ is reduced. This entails increased profitability which invites more entry by our assumption (Equation (6.2)). Equations (6.1) and (6.4) comprise the investment model. This model considers investment as the driving force of net capital accumulation. While these three models are broadly similar, their stability characteristics measured by the eigenvalues are very different.

6.3 Evolution under the cost model

Consider the industry evolution under the cost model defined by Equations (6.1) and (6.2). The dynamic problem for the incumbent firm is to select the time path of the R&D expenditure $u(t)$ such that it minimizes the initial discounted cost $C_0 = C(0)$ subject to the state Equations (6.1) and (6.2). The current value Hamiltonian is

$$H = c(y,u) + s_1(a(p - c(y,u))) + s_2(rC - c(y,u)) \tag{6.5}$$

where s_1 and s_2 are the respective co-state variables associated with Equations (6.1) and (6.2). Clearly the optimal control function minimizes the expression $(1 - s_2)c(y,u)$ and by Equation (6.2) the optimal control is $c(y)$. The co-state variable s_1 is the shadow price of the new firm and it is positive for all t if its initial value $s_1(0)$ is chosen in a proper way, since

$$\dot{s}_1 = rs_1 - (1 - s_2)\, c'(y), c'(y) = \frac{\partial c}{\partial y}$$

Note that the optimal control does not depend on the co-state variables, so that we may ignore the co-state variables and focus on the dynamics defined by the state variables y and C. This is the system

$$\dot{y} = a(p - c(y)) \tag{6.6}$$

$$\dot{C} = rC - c(y) \tag{6.7}$$

When $p = c(y)$ there is no new entry into (or exits from) the industry in the long run. The steady state output y^* and the associated number n^* of firms in the industry are given by the solution of the equation

$$p = c(y) = rC \tag{6.8}$$

Clearly the shape of the average cost function $c(y)$ determines the nature of evolution of firms in the industry. The steady state cost C^* is given by

$$C^* = \frac{c(y)}{r}$$

On linearizing the dynamic system (6.6) and (6.7) the characteristic equation may be written as

$$\lambda^2 + (ac' - r)\lambda - rac' = 0$$

with eigenvalues

$$\lambda = \left(\frac{1}{2}\right)\{(r - ac') \pm [(r - ac')^2 + 4rac']^{1/2}\} \tag{6.9}$$

Several cases may be analyzed. In case $c' = \partial c / \partial y$ is positive and the eigenvalues are real, then one eigenvalue is positive and one negative. This yields the stationary state to be a saddle point. When c' is negative and a is sufficiently small, both characteristic roots are positive and the stationary equilibrium is an unstable node. Finally, if c' is negative and $(r - ac')^2 < 4rac'$, then the roots are complex-valued with positive real parts. In this case the stationary state is an unstable focus, where cyclical evolutions persist. Thus the equilibrium path of industry evolution is a trajectory of the dynamic system (6.6) and (6.7) around any of the three long-run equilibria. Given the initial number of firms n_0 with associated output y_0, the equilibrium path is determined by the initial cost C_0. The cyclical dynamics imply a multiplicity of equilibrium paths. For a given initial number of firms n_0 with output y_0, there may be many initial values C_0 specifying the movement towards one of the steady states.

In case of saddle point stationary state, a stable manifold along which motion is purely towards (y^*, C^*) and an unstable manifold along which motion is exclusively away from (y^*, C^*) are given by the eigenvectors of the coefficient matrix of the system (Equations (6.6) and (6.7)) in a linearized form corresponding to the stable and the unstable roots respectively.

6.4 Evolution under the demand model

The dynamic evolution specified by the demand model given by Equations (6.1) and (6.3) follows the Walrasian adjustment process, where excess profits tend to increase output over time and excess demand tends to increase the market price over time. The equilibrium stationary state is now given by (p^*, y^*) when $\dot{y} = 0 = \dot{p}$. On linearizing the system around the equilibrium point (p^*, y^*) one obtains the matrix system

$$\begin{pmatrix} \dot{y} \\ \dot{p} \end{pmatrix} = \begin{bmatrix} -ac' & a \\ -b & bD' \end{bmatrix} \begin{pmatrix} y \\ p \end{pmatrix} \tag{6.10}$$

This has the quadratic characteristic equation

$$\lambda^2 + (ac' - bD')\lambda + ab(1 - c'D') = 0 \tag{6.11}$$

with two roots. Both roots have negative real parts if and only if $ac' - bD' > 0$ and $ab(1 - c'D') > 0$. But since the demand curve has a negative slope around (y^*, p^*), we have $D' < 0$. Hence if $c' > 0$, that is increasing average cost, then the above conditions hold and hence each root has a negative real part. This shows that (y^*, p^*) specifies a globally stable equilibrium in this situation.

If, however, average cost declines, that is $c' < 0$ due to higher R&D investments, then the situation may become different. For example, if increasing returns to scale are sufficiently large so that the following inequality

$$\frac{c' < 1}{D' < 0} \tag{6.12}$$

holds, then the two roots may specify a saddle point. For example the roots computed from Equation (6.11) are given by

$$\lambda = \left(\frac{1}{2}\right)(ac' - bD') \pm \left(\frac{1}{2}\right)[(ac' - bD')^2 - 4ab(1 - c'D')]^{1/2} \tag{6.13}$$

Now if Equation (6.12) holds, that is $c'D' > 1$, then we have two real solutions in λ, one positive and one negative. The equilibrium point (y^*, p^*) then specifies a saddle point. Hence there is a stable manifold along which the industry evolution is purely towards (y^*, p^*) and an unstable manifold along which the evolution is away from the equilibrium point. The slopes of these manifolds at (y^*, p^*) are given by the eigenvectors of the coefficient matrix in Equation (6.1) corresponding to the stable and unstable roots respectively. Note that the condition $c'(y) < 0$ represents increasing returns and the major source of this is productivity gains through innovations as denoted by u in Equation (6.2). On taking the time derivative of both sides in (6.1) we obtain

$$\ddot{y} = a(\dot{p} - c'\dot{y}) \tag{6.14}$$

where $\dot{y} = \mathrm{d}y/\mathrm{d}t$ is positive along the unstable manifold and hence $-c'\dot{y}$ is positive for $c' < 0$. But $\dot{p} = b[D(p) - y]$, which is nearly zero near the graph of the demand function $D(p)$ which coincides with the unstable manifold near the equilibrium, that is the gap $[D(p) - y]$ is nearly zero on the stable manifold near the equilibrium point. Heal (1986) has analyzed this system as a macroeconomic adjustment process and shown that the concept of convergence used here is different. It means that in the long run the system remains in a region of its phase space within which certain properties hold, and not that it actually converges to a point. Thus one may conclude that if there are increasing returns in production due to innovations, the dynamic economy moves toward one of two regimes. In one the outputs and profits rise and prices fall. The other exhibits falling profits and outputs and rising prices. The apparent existence of "vicious" and "virtuous" circles of industry evolution suggests the need for a rationally public sector policy toward fiscal controls.

An interesting extension of the demand-oriented Walrasian adjustment model has been discussed by Goodwin (1989) in terms of feedback control theory. On interpreting the demand model as a description of a proportional autocontrol mechanism operating through cross-effects between prices and outputs, he introduced derivative control rules, where supply of output reacts not solely to differentials in price–cost margins but also to the rate of change of those differentials. Similarly price adjustments are not only a sign-preserving function of excess demand but prices also respond to the rate of change in excess demand. Thus

$$(p-c)+\Delta(p-c) \rightarrow \dot{y} \tag{6.15}$$
$$(d-y)+\Delta(d-y) \rightarrow \dot{p}$$

where Δ means the time rate of change of excess demand or excess of price over cost. Goodwin has shown that the introduction of such derivative control mechanisms enhances the stability properties of the dynamic evolution significantly. In addition the derivative control mechanism may help us to understand why the price output fluctuations may be globally bounded, though not necessarily convergent to one steady state.

On linearizing the derivative control system (Equation (6.15)) as before, the transformed system can be formalized as

$$\dot{y} = \alpha_1 p - \alpha_2 y \tag{6.16}$$
$$\dot{p} = \beta_1 p - \beta_2 y$$

where

$$\alpha_1 = (1 - a_2 p' + a_2 c')^{-1} a_1$$
$$\alpha_2 = a_1 c'(1 - a_2 p' + a_2 c')^{-1}$$
$$\beta_1 = b_1 d'(1 - b_2 d' + b_2 y')^{-1}$$
$$\beta_2 = b_1 (1 - b_2 d' + b_2 y')^{-1}$$

and

$$\dot{y} = a_1(p-c) + a_2(p'\dot{y} - c'\dot{y}) \tag{6.17}$$
$$\dot{p} = b_1(d-y) + b_2(d'\dot{p} - y'\dot{p})$$

The characteristic roots of Equation (6.16) are

$$\lambda = \left(\frac{1}{2}\right)[\beta_1 - \alpha_2 \pm (\alpha_2 - \beta_1)^2 + 4(\alpha_2\beta_1 - \alpha_1\beta_2)^{1/2}]$$

Equations (6.16) and (6.17) state that output supply is adjusted following the discrepancy between the current price and the marginal cost, and the rate of change of this discrepancy and prices are adjusted in the direction of the excess demand on the market for goods and the rate of change of this excess demand. Clearly the derivative control may speed up the adjustment or evolution process. Thus the concept of a derivative autocontrol may not always be applicable in the vicinity of steady state due to the existence of positive real parts of roots but it may help explain why the price-output fluctuations may remain globally bounded, though not necessarily convergent to a steady state.

6.5 Evolution under the investment model

The investment model assumes that average costs may be reduced by firms choosing higher scale measured by the net capital stock, which increases capacity output. This capital stock may be primarily in the form of knowledge capital, for example R&D or innovations. This model may be rewritten in terms of average industry cost $\bar{c}$ and its deviation from the innovating firms' cost $c(k)$, that is

$$\begin{aligned} \dot{y} &= a(\bar{c} - c(k)) \\ \dot{k} &= I - \delta k \end{aligned} \tag{6.18}$$

Thus firms grow in size if $c(k)$ is less than $\bar{c}$. By improving cost-efficiency a firm can grow faster. This type of model is closely related to the dynamic evolution model developed by Mazzacato (2000) and Metcalfe (1994), who argued that the cost reduction process, also called the dynamic increasing returns, may occur at diverse rates for different firms thus increasing the comparative advantages of the successful firms and decreasing the same for the laggards. Two interesting implications follow from this type of evolution model. First, the major source of growth here is the productivity gain or efficiency. Any means which help improve efficiency would improve the growth in size measured by output. This has been empirically supported by several studies; for example, Lansbury and Mayes (1996) have found for industrial data in the United Kingdom that the entry and exit processes are mainly explained by the rise and the fall of productive efficiency respectively. Secondly, this provides the basis of the modern evolutionary theory of competition, which has borrowed some key ideas of Fisher (1930) and his competitive fitness model of growth of biological species. In the Fisherian model the replicator dynamics in evolutionary biology is formalized as

$$\dot{x}_i = Ax_i(E_i - \overline{E}), \quad \overline{E} = \sum_{i=1}^{n} x_i \, E_i$$

where x_i is the proportion of species i in a population, E_i is its reproductive fitness and $\overline{E}$ is the mean fitness. Fisher's fundamental theorem in replicator dynamics states,

$$\frac{\mathrm{d}\overline{E}}{\mathrm{d}t} = -\alpha V, \quad \alpha > 0$$

That is, the rate of change in mean fitness is inversely proportional to the variance of fitness characteristics in the population. In terms of the competitive market dynamics in Equation (6.18) the mean fitness would be measured by the negative of $\bar{c}$ and hence

$$\frac{\mathrm{d}\bar{c}}{\mathrm{d}t} = \alpha V$$

Thus the output evolution equation (Equation (6.18)) can be written as

$$\ddot{y} = a(\alpha V - \bar{c}(k)) \tag{6.19}$$

On linearizing one obtains the system

$$\begin{aligned} \ddot{y} &= a(\alpha V - c'\dot{k} - c_0) \\ \dot{k} &= I - \delta k \end{aligned} \tag{6.20}$$

Note that $c' = \delta c/\delta k$ is negative when there is dynamic increasing returns or cost economies due to innovations. If $c' > 0$ then investment $I(t)$ involves more costs and hence it lowers the acceleration of output.

Thus for any given rate of net capital accumulation ($\dot{k}$), the higher the variance of average costs i in any industry, the higher the acceleration of output. Similarly a higher rate of growth of net capital ($\dot{k}$) helps to increase the rate of acceleration of output. Denoting the acceleration of output as $A = \ddot{y}$, the evolution equation (Equation (6.20)) can also be written as

$$A = a_0 + a_1 V - a_2 k + a_3 I \tag{6.21}$$

where $a_0 = -ac_0$, $a_1 = \alpha x$, $a_2 = -ac'\delta$, $a_3 = -ac'$. Since a_3 is positive under dynamic increasing returns, the higher the investment flow (I), the higher the acceleration of industry output. If the depreciation rate δ is ignored, one obtains the two major sources of acceleration of output, that is variance and investment.

When average cost is viewed as a function of output and capital stock, then the differential Equation (6.19) may be written as

$$\ddot{y} = a(\alpha V - c'\dot{k} - c_y\dot{y} - c_0), \; c_y = \frac{\delta c}{\delta y} \tag{6.22}$$

This has the characteristic equation

$$\lambda^2 + \beta_1\lambda - \beta_2 = 0$$

where $\beta_1 = ac_y$, $\beta_2 = a\alpha V + a_3 I$ when $\delta = 0$
The two roots are

$$\lambda_{1,2} = \left(\frac{1}{2}\right)(-\beta_1 \pm (\beta_1^2 + 4\beta_2)^{1/2})$$

The roots are real since β_2 is positive and they are of opposite sign. Hence there is a saddle point equilibrium. The slope of the unstable manifold is given by the eigenvectors of the matrix associated with the differential Equation (6.22).

An advantage of the formulation (6.21) of the evolution process is that the stochasticity of the investment process can be easily built into it. For example Dixit and Pindyck (1994) have developed a geometric Brownian model for the capital accumulation process. Gort and Konakayama (1982) have applied a diffusion model in the production of an innovation through knowledge capital. Following this procedure assume that there is no depreciation so that $\dot{k} = I$ and let investment (I) grow as

$$\frac{dI}{dt} = r(t)\, I(t) \tag{6.23}$$

where the rate of growth $r(t)$ follows a random process

$$r(t) = g + h\, u(t)$$

where $u(t)$ is a white noise Gaussian process. The stochastic differential Equation (6.23) becomes

$$dI(t) = I(t)[g\, dt + h\, dB(t)], \quad u(t)\, dt = dB(t) \tag{6.24}$$

The solution of this equation in terms of the Ito integral produces a diffusion process characterized by the drift ($\mu(x)$) and variance ($\sigma^2(x)$) coefficients as $\mu(x) = gx$ and $\sigma^2(x) = h^2x^2$. Since the solution of Equations (6.22) and (6.23) is of the form

$$I(t) = I(0)\ \exp[gt + h\ B(t)]$$

it is clear that whenever $0 < g < h^2/2$, $I(t)$ tends to zero as $t \to \infty$. Otherwise $I(t)$ may rise over time. Thus stochasticity may provide additional sources of growth or decline of an industry over time.

Sengupta (2004) has analyzed the investment model to explain the market dynamics of the evolution of industry and the impact of new technology with R&D and the knowledge capital.

6.6 An application in the computer industry

Competition and technological change has been most rapid in the computer industry today and their impact on productivity growth has been most significant over the last two decades. Thus Norsworthy and Jang (1992) have empirically found for the US computer industry a productivity growth rate of 2 percent per year for the period 1958–1996, while for the period 1992–1998 the growth rate exceeded 2.5 percent per year on the average. Increased R&D investment and expanding "knowledge capital" have made a significant contribution to this productivity growth.

They helped reduce average costs through significant economies of scale and learning curve effects.

The computer industry is in many ways unique among the high-tech industries of today in two respects. First, its efficiency gains have large spillover effects on other industries, for example electronics, telecommunications, transportation and so on. Second, the technological progress which has been emphasized by Solow (1997) and Lucas (1993) in modern growth theory has been most rapid in this industry through research in software development and so on.

The scale economies and efficiency in this industry are estimated here by a nonparametric approach called "data envelopment analysis" (DEA). This method is more robust than the nonlinear maximum likelihood approach adopted by Norsworthy and Jang (1992). This is because the DEA method is based on a nonparametric approach using the technique of LP.

The cost function used here is the modified form of a translog function

$$\ln\ TC = b_0 + b_1\ \ln\ Y(t) + b_2 t$$

where TC = total cost, $Y(t)$ = output and t is a time trend variable used as a dummy for the state of technology as in the Solow model. On differentiating with respect to time one obtains

$$\frac{\Delta \text{TC}}{\text{TC}} = b_1 \left(\frac{\Delta Y}{Y(t)} \right) + b_2$$

We replace the time proxy by R&D investment (R) and the cost frontier is then reduced to

$$\frac{\Delta \text{TC}}{\text{TC}} = b_1 \left(\frac{\Delta Y}{Y(t)} \right) + b_2 \left(\frac{\Delta R}{R} \right)$$

The following DEA model is then set up to estimate the cost frontier

$$\begin{aligned} &\text{Min } \theta \\ &\text{subject to } \sum_{j=1}^{n} c_j \lambda_j \leq \theta c_h \\ &\sum_{j=1}^{n} y_j \lambda_j \geq y_h \\ &\sum_{j=1}^{n} r_j \lambda_j \leq r_h \\ &\sum_{j=1}^{n} \lambda_j = 1; \lambda_j \geq 0; j = 1, 2, \ldots, n \end{aligned} \tag{6.25}$$

where the reference unit h is tested if it is on the cost frontier defined by the efficient units in the industry. Here $c_j = \Delta \text{TC}_j / \text{TC}_j$, $y_j = \Delta Y_j / Y_j$, and $r_j = \Delta R_j / R_j$. On maximizing the Lagrangean function L

$$L = -\theta + \beta_1 \left(\theta c_h - \sum_{j=1}^{n} c_j \lambda_j \right) + \beta_2 \left(r_h - \sum_{j=1}^{n} r_j \lambda_j \right) + \alpha \left(\sum_{j=1}^{n} y_j \lambda_j - y_h \right) + \beta_0 \left(\sum_{j=1}^{n} \lambda_j - 1 \right)$$

One obtains the dynamic cost frontier

$$c_h^* = \gamma_0 + \gamma_1 y_h^* - \gamma_2 r_h^* \tag{6.26}$$

if unit h is efficient, that is $\theta^* = 1$, $c_h = \Sigma c_j \lambda_j^*$ and $\Sigma y_j \lambda_j^*$ and $\gamma_0 = \beta_0/\beta_1$, $\gamma_1 = \alpha/\beta_1$, $\gamma_2 = \beta_2/\beta_1$. If unit h is not efficient then

$$\sum_{j=1}^{n} c_j \lambda_j^* < c_h$$

which implies cost-inefficiency, since other units (or firms) in the industry can do better. The set of firms ($h = 1, 2, \ldots, n_1$; $n_1 \leq n$) which is on the dynamic cost frontier (Equation (6.25)) has two differentiating characteristics from the inefficient firms. First, the R&D inputs help reduce the rate of change in costs since γ_2 is positive. Indirectly this measures the productivity impact due to dynamic increasing returns and learning by doing. Secondly, a negative value of γ_0 indicates a downward shift of the cost frontier indicating cost-efficiency. Since the Lagrange multiplier β_0 is free in sign, the sign of γ_0 may be either positive or negative. Thus the cost frontier model (Equation (6.25)) exhibits efficiency gains due to time shift and the growth of R&D investment.

Two implications of the DEA model (Equation (6.25)) have to be noted. First, the output growth by efficient firms can be significantly influenced by the growth of R&D inputs and related innovations. Secondly, the dynamic production frontier implicit in the cost frontier Equation (6.25) of the efficient firms can exhibit significant increasing returns if the ratio β_1/α exceed unity, that is γ_1 is less than unity. Thus R&D investment helps to increase output and also reduce the rate of cost increase.

The empirical application of the DEA model (Equation (6.25)) to the computer industry over the period 1985–2000 has been discussed by Sengupta (2004) in some detail elsewhere. The impact of R&D inputs (r_j) measured by the coefficient γ_2 in Equation (6.25) will be discussed here.

The empirical data are obtained from Standard and Poor's Compustat Database (SIC Codes 3570 and 3571) and a set of 10 companies are selected out of a total of 42. This selection is based on two considerations: (i) survival for the whole period and (ii) promising profit records. The choice of a three-year average subperiod is motivated by the fact that this is usually the time of development of new product vintages in the technological development in the computer industry.

Note also that the R&D expenses here include both software development and related research expenses and marketing innovations. Data limitations prevent us from considering only the research-based expenses. Two interesting points emerge from Table 6.1. First, the high degree of elasticity ($\gamma_2 > 1$) of costs with respect to R&D inputs whenever

Table 6.1 Impact of R&D inputs on cost-efficiency

	1985–1989		1990–1994		1995–2000	
	θ^*	γ_2	θ^*	γ_2	θ^*	γ_2
Dell	1.00	2.71	1.00	0.15	0.75	0.08
Compaq	0.97	0.03	1.00	0.15	0.95	0.001
HP	1.00	1.89	0.93	0.10	0.88	0.002
Sun	1.00	0.001	1.00	0.13	0.97	1.79
Toshiba	0.93	1.56	1.00	0.12	0.82	0.09
Silicon Graphics	0.99	0.02	0.95	1.41	0.87	0.001
Sequent	0.72	0.80	0.92	0.001	0.84	0.002
Hitachi	0.88	0.07	0.98	0.21	0.55	0.001
Apple	1.00	1.21	0.87	0.92	0.68	0.001
Data General	0.90	0.92	0.62	0.54	0.81	0.65

firms are dynamically efficient (i.e., $\theta^* = 1.0$). Second, the efficiency score (θ^*) has changed over the years due to intense competition and market uncertainty. This implies that different types of stability or instability may hold in different periods. The theoretical models of stability/instability analyzed before may thus become relevant in some subperiod.

6.7 Conclusion

Profitability and R&D investment are the major determinants of industry evolution today in the high-technology fields. A set of dynamic models is developed and analyzed here in terms of their stability and instability characteristics. Although the saddle point equilibria are shown to be more likely in this evolutionary framework, the implications of unstable manifolds are also discussed. An empirical application in the computer industry is that the impact of R&D investment on cost-efficiency has been significant for firms which remained close to the dynamically efficient cost frontier.

References

D'Aveni, R.A. (1994) *Hypercompetition: Managing the Dynamics of Strategic Maneuvering* (New York: Free Press).

Dixit, A. and R. Pindyck (1994) *Investment Under Uncertainty* (Princeton: Princeton University Press).

Fisher, R.A. (1930) *The Genetical Theory of Natural Selection* (Oxford: Clarendon Press).

Goodwin, R.M. (1989) *Essays in Nonlinear Economic Dynamics* (Frankfurt: Verlag Peter Lang).

Gort, M. and A. Konakayama (1982) A model of diffusion in the production of an innovation, *American Economic Review* 72, 1111–1120.

Heal, G. (1986) Macrodynamics and returns to scale, *Economic Journal* 96, 191–198.

Lansbury, M. and D. Mayes (1996) Entry, exit, ownership and the growth of productivity, in D. Mayes (ed.) *Sources of Productivity Growth* (Cambridge: Cambridge University Press).

Lucas, R.E. (1993) Making a miracle, *Econometrica* 61, 251–272.

Mazzacato, M. (2000) *Firm Size, Innovation and Market* (Cheltenham: Edward Elgar).

Metcalfe, J.S. (1994) Competition, evolution and the capital market, *Metroeconomica* 4, 127–14.

Norsworthy, J.R. and S.L. Jang (1992) *Empirical Measurement and Analysis of Productivity and Technological Change* (Amsterdam: North Holland).

Sengupta, J.K. (2004) *Competition and Growth: Innovations and Selection in Industry Evolution* (New York: Palgrave Macmillan).

Solow, R.M. (1997) *Learning from Learning by Doing: Lessons for Economic Growth* (Stanford: Stanford University Press).

Index

0 1341 1049091 6

RECEIVED